WORLD OF FURS

WORLD OF FURS

David G. Kaplan

edited by Ed Kleinman

Fairchild Publications, Inc.
New York

Standard Book Number: 87005-098-2

Library of Congress Catalog Card Number: 75–153567

Designer: Dick Kluga

Printed in the United States of America

TABLE OF CONTENTS

Introduction: Industry in Perspective

Long before Man had a lust for gold or diamonds, he was panting—and even on occasion dying—for the natural coat of animals.

Since the first primitives discovered the warmth of animal skins, people have spent untold millions of dollars to drape themselves in furs that have come to symbolize wealth, status, love, power, beauty and good taste.

Competition for the best of fur skins has extended to palaces and Presidential suites, and some of the history of this ancient industry has been written in high adventure, intrigue, tragedy, and even blood.

Kings and queens alike have been known to put great weight in wearing the best of ermine and mink at courtly affairs. Trappers in the coldest reaches of Russia have died in quest of the choicest sable skins. In contemporary times, the entire mink industry has been stood on its embarrassed head by a First Lady.

Each fur season has at least one great auction scene where the key dealers, manufacturers and retailers gather to do battle over a new mink mutation, a particularly silky lot of chinchilla, the richest, fullest-furred collection of Russian sable.

The dramatic auction event reflects the real dynamics of this industry. There is the tension as the great purchasers enter the auction hall; the gambling spirit about whether a new color will succeed; the secret bidding signals that the "aficionados" will try to spot; the staccato of the auctioneer, as prices move upward, and as the struggle narrows down to a handful of the fittest.

I covered this scene for seven years as fur editor for the leading business paper, *Women's Wear Daily*. During that period I discovered the limitations of fur industry literature. While there were several first-rate books, none covered the full picture, none put the industry in historical perspective.

When I became a book publisher several years ago, I wondered whether a first-rate reference book could be produced to fill this void.

Enter Dave Kaplan. For many years I had known and respected him in his various roles in the industry: columnist, author, teacher, businessman. He had spiritually wrapped himself in fur industry history. As a researcher he proved unrelenting. To round out an editorial team, we asked Ed Kleinman, consultant, educator and head of the Fur Department at the H.S. of Fashion Industries in New York, a generation younger than Dave but just as devoted to the world of furs, to serve in an editing capacity. He checked out the massive material, helped select and caption illustrations, brought the manuscript up-to-the-minute, and placed the ecological impact on the industry in perspective.

All of the people involved in this project shared a common purpose. This book was prepared to meet a need in several forums:

First, there was a commitment to business history. The book was produced to provide school and public libraries with the first authoritative record on every important aspect of this historical industry.

Second, the book would be of considerable interest to business and professional people in related fields—apparel manufacturers, designers and fashion writers, for example—whose careers inevitably touch the fur industry.

Third, the book was conceived to be a career stimuli for thousands of students and industry trainees who might discover a niche for their skills and desires.

Finally, and unashamedly, the book was written for people in the fur industry. It is, after all, the record of an extraordinary business, an industry marked by paradox, tension, glamour, humor, creativity, genius and tragedy. For those who have been part of this story, the book will be something of a sentimental journey, a fitting aspect of a very human experience.

E. B. Gold, Mgr.
FAIRCHILD BOOK DIVISION

Acknowledgments

Many leaders of the fur industry made extraordinary contributions to this book. First on the list would be Jess Chernak, Executive Secretary, Fur Information and Fashion Council, who opened up FIFC's extensive photo collection for our use, and also provided us with a great deal of literature, particularly on contemporary issues.

The Fairchild Photography Department also took a substantial number of original pictures at these fur companies: Admiration Fur Dyeing Corp., Alexion and Son, Antonivich Bros. Inc., Chatman-Fuhrman Inc., Christie Bros., J. Corn and Son Inc., Hudson's Bay Fur Co., J. Kasindorf Inc., Sidney Lambert Inc., Marco Royal Dark Inc., Quality Manos Block Inc., Schreibman-Raphael Inc., A. Winick and Son, U.S. Cleaners Inc. and Valerie Furs Corp.

Others who deserve mention include: Dick Kluga, who designed the book; Fred Romary and Henry Fera, artists, who helped with the historic pictures; and Donna Fyler and Marjorie Lewis, of the Fairchild Books & Visuals editorial staff.

1

The Power and Prestige of Furs

1. The Power and Prestige of Furs

Jason searched for the Golden Fleece and found a lamb skin. That lamb skin was to Jason and the Greeks what the discovery of gold was to this country in the 1800's. In ancient times furs meant wealth, power, and prestige. The classic Greeks portrayed their gods dressed in skins. The ancient city of Tyre thought that the discovery of the fur dyeing power of the porpura shell (by Hercules' dog, it is said) was important enough to mint a commemorative coin.

Jason, however, was a latecomer to the world of fur. Man has used and prized furs from his earliest beginning. Food, covering, shelter, ornaments, and tools are among the uses our cave-dwelling ancestors made of animals. Today we still make the same uses of both wild and domestic animals. For leather, fur, ornamentation, and decoration we still turn to fur-bearing animals. Through necessity and desire the world of fur has from the beginning been intricately woven into our culture, religions, customs, mythology, laws, ecology, and economy.

In ancient days, some curious customs, beliefs, and ideas grew up around furs. The Greeks considered the wearing of skins as barbaric and used furs only as rugs or furniture coverings. Roman fur-

riers, on the other hand, assured the public that seal and hyena skins were protection from lightning, while deerskins were a safeguard against snake bites. Roman medical practitioners prescribed beaver skins for burns and scalds and beaver-skin shoes for gout. A single sheepskin covering two chairs upon which bride and groom sat was an integral part of a Roman wedding ceremony. Queen Sem-

In the beginning: when people were still living in caves, furs were important for their warmth.

iramis returned in 21 B.C. from India with 8000 tiger skins to decorate the floors of the hanging gardens of Babylon. In 1973 several well known New York City department stores promoted furs as a "new" decoration for the modern bedroom and living room.

The Middle Kingdom of Egypt reserved the use of leopard skins for royalty, a policy followed later by many African tribes and kingdoms. At various times and in various countries such furs as sable, ermine, chinchilla, and Alaska seal, have been restricted to royalty. Edward I of England forbade the common woman to wear a hood trimmed with any fur other than lamb or rabbit. Philip the Good of France prohibited bourgeois women from wearing mink, squirrel, or ermine. The Incas made a royal fur of chinchilla and the royalty of Western Europe made ermine their private symbol of authority. The clergy used the white ermine as a sign of purity, and perhaps through suggestion, a challenge to the power of the throne.

On the other hand, the Roman city government of the Empire days, probably reacting to the clothing worn by its enemies, banned the use of furs in the center of the city, calling them barbaric and smelly. Fortunately, the Romans did not

Plains Indians used furskins for a variety of necessities: tents, rugs, blankets. This activity about 1800.

have the last word. Today furs are still a sign of status, authority, and wealth. A woman's furs symbolize milady's place in society. Her ego and prestige are boosted if she can wear mink or sable in a gathering of rabbit or lamb.

Furriers and the fur trade entrepreneurs were among the most prominent citizens of their day. Cardinal Richelieu, the power behind the throne of two Louis, founded "La companie. . . . . .Pelleteries," making himself a director. Prince Rupert of England granted a charter to the Hudson's Bay Company in 1670 to get the valuable beaver skins from Hudson Bay in Canada. Cardinal Richelieu and his *companie* are no more, but the Hudson's Bay Company is still in business more than 300 years later. It has 214 trading posts in Canada reaching to the almost inaccessible North Pole and still plays an important part in the Canadian economy.

The fur trade and the lure of wealth in the form of beaver started the settlement of the Americas. The Astor fortune was founded on the beaver trade. By 1780 Astor had absorbed ten competitors into the American Fur Company, one of the first American "conglomerates." Beaver was the first form of "money" in the early colonies. Indeed, two beavers are to be found on the official seal of the state of New York. Pierre Laclède, who founded New Orleans, ranged as far as Lakes Michigan and Machinac and Quebec, trading and shipping furs.

Today the trapper and hunter play a much smaller part in the world of fur. Wild animal areas have decreased and the business of raising animals for their pelts has grown in proportion. The number of furs ranched are legion. Many have been transported to new areas and released to run wild, or semi-controlled in a preserve.

It is this story of furs, tied to the history and background of Man, that we now pursue. Man cannot be separated from the animals and nature. The life of one is the life of the other. Ours is, indeed, a World of Furs.

Much of the historical information in this chapter was excerpted from:

Francis Weiss, *Adam to Madam* (London, *International Fur Review*. Excerpts printed irregularly, 1963–1967).

Dr. Eva Neinholdt, *Furs in European Clothing* (Leipzig. Hermalon Series. Excerpts printed irregularly, 1964–1968).

1 "Furs and Their Romantic Past," *Fur Age Weekly*, March, 1971, p. 3.

2

Fur Fashion History: Something for Everyone

2. Fur Fashion History: Something for Everyone

THE EARLIEST FUR GARMENTS

Originally, skins, pelts or hides were probably worn in their natural shapes, with the feet used to tie the pelt over the shoulder or around the waist. Male and female Hebrews in Old Testament days wore a garment known as the *kaunaki*, made from a shaggy lamb hide that covered them from the waist down.

Early records show the Nile region dwellers wearing sheepskins in much the same way as do present Outer Mongolians. This is one of the earliest types of fur garment, consisting of a wide-armed fur cross with a head-sized hole in the middle. The wearer slips it over his head. The shorter wings become the arms and the longer sections cover the front and back.

One of the earliest "designs" involved the cutting and trimming of fur to a desired form called the *Pelisse*, a long rectangle of fur with an opening for the head and with the long sections hanging from the front and back. This development paralleled or preceded the Roman toga, which is basically the same structure refined by a shoulder fastening and some form of tie or belt. Romans, in general, did not wear furs until about the 3rd Century, but the pioneer Roman Senators were called "Pelliti" because they did wear pelts. At the height of the Byzantine Empire we find furs used as trimmings.

FURS IN THE MIDDLE AGES

The historical high point in the use of furs were the centuries between the fall of Rome and the Renaissance. A garment lined or trimmed with fur was a necessity in those days of fireplace-heated homes, drafty halls, and snow-covered dirt roads. Travellers, crusaders, and explorers were followed by traders who brought back new furs and new ideas for their use. Marco Polo was not believed when he described the immense tents of Genghis Khan, large enough to hold thousands and covered with skins so well joined that they shed rain and snow.

As early as the 8th Century, the pelisse was lined or trimmed with fur. The Gothic man might wear a half circular cape or a cape with a fur lining and border, open in front and fastened over the shoulder. The omnipresent sleeveless cloak acquired a fur lining around the 12th Century. France's King Louis IX limited himself to a "pelisson" robe lined in unpretentious black lamb, but most princes maintained no such restraints, and often lined their floor-length cloaks with ermine.

The ancestor of the coat, the *coat-hardie*, appeared in the 13th Century. Close-fitting, with tight sleeves bordered in fur, it ranged from calf-length, upwards to the thighs, and was liberally trimmed with fur on sleeve and sweep edges. Women wore a variation of this garment, sometimes covered with a fur stomacher. It was combined with long-hanging, fur-trimmed "angel" sleeves for the nobility. Women of the French Court added jewels to their ermine-trimmed versions, which was tightly fitted, cut away in front, and extremely elaborate. A century later what was unmistakably a jacket, velvet with fur trim, became common in European Court, paintings and records, usually worn under a fur-trimmed mantle.

In the 14th Century Edward III of England made ermine, spotted with symmetrical black spots of lamb, a royal fur. The long cloak of the subsequent style was decorated with fur facings and cuffs which differed from the fur used for the lining.

As the cord-held cloak retired from the fashion of the 14th Century, it was replaced by an open mantlet with turned-wide collar. Another version of the cloak appeared, of voluminous dimensions, all-enveloping, floor length, and far more elaborate and drapable than its predecessor. The *houppelande*, as it was called, was collared and lined with fur. Nobles used sable ermine and marten, while commoners contented themselves with squirrel or lamb. More formal versions, which swept the floor and had matching long sleeves lined with fur, became popular with women.

The more utilitarian upper garments then popular were called the *succene* and the *surcoat*. The succene was a sleeveless long vest with slits for the arms, worn over a skirt or dress. The surcoat had deep half-sleeves slit half way up for ease of movement, a feature important when worn over armor. The slits showed the fur lining underneath. The version of this garment for women, the *chasuble dress*, also sported a fur edging or lining. The idea of a changeable lining (from fur to silk) to conform with the seasons had already been developed. Fur mittens, fur-lined shoes, and blankets of fur were common at all levels of society, even though the kind of fur which one could use was often regulated by law to conform with social position. These laws extended even to millinery. The peasant sported sheepskin while the nobility wore ermine. The forerunner of the modern ski-cap and parka hood was the 14th Century *gugel*, a hood and collar combination worked as one, with the fur lining showing. The Germanic people had their version of this item with a flared standing collar, the *coller*, which has persisted to the present day in national costumes of the area. Fifteenth Century Florentines called their version of the hood and cape unit a *chaperone*.

Early clerical regulations limited the wearing of luxury furs, in most instances restricting the lower clergy to sheepskins.[1] Upper levels of the clerical hierarchy, including Popes, wore the *cappa magna*, a long trailing coat with a combination collar-cowl, lined with gray squirrel or ermine. The shorter *mozzetta*, a front-buttoned ceremonial cape, was and still is restricted to the very highest clergy. The mozzetta given to Pope Paul VI was of purple velvet lined in white mink. Nuns of at least two orders wore cloth coats trimmed with ermine and often lined with rabbit.

FURS IN THE RENAISSANCE

The flow of Russian furs, which distinguished the era of the Crusades, was augmented by the discovery of the New World and its immense resources of furs. By the 16th Century, the Spanish *cotte* or coat, fur-lined and fur-trimmed, was universally fashionable in Europe. The large fur muff not only kept hands warm but also served as a symbol of social status. The fur scarf made its first appearance held in place by elaborately jewelled pins and chains. Nor did the eastern peoples neglect furs in their costumes. When the Turks captured Constantinople in 1453, they also adopted the Byzantine taste for elaborately furred dress. Their caftans and robes were liberally trimmed with pelts of all kinds.

The wide-shouldered, puff-sleeved gown associated with Henry VIII was by no means exclusively his. All over Europe, brocaded, fur-lined, and fur-trimmed knee-length gowns were the stylish Court outfit of nobility and courtières. The fur scarf called a *tippet* was a common decorative accessory over a dress, around the neck, over a shoulder, or hung from the waist.

The ankle-length robe was by this time the traditional official dress for men of social standing. It was invariably trimmed with fur on all the edges in the same way capes and collars were trimmed. Women's costumes also included wide fur-banded sleeves, or false sleeves entirely of fur. How styles repeat themselves! In 1972 the fur sleeve has reappeared, this time on a cloth jacket.

A very fancy fur coat made for this Eastern European bride, about 1600. Checkered effect was unique then.

By the 16th Century, the accepted male attire was the doublet and hose. Over this, the well-dressed male wore a cloth, fur-lined cape, short enough to leave his sword free. These full capes were elaborately trimmed and decorated. The fur collar extended full length down the front. About this time, occurred the first attempts to make a fully furred bridal outfit with the fur outside, a style previously left to the peasantry. In the 1974–1975 fur season, no bride would think of wearing a bridal gown of fur, but we have returned to this style in reverse by bringing back the fur coat with the fur on the inside and the leather on the outside. This is seen today in the shearling lamb coats and jackets held together with buttons, frogs, and chains. They are often trimmed with cuffs, collars, and borders of contrasting fur.

17TH CENTURY

Anyone who could afford it and wished to be regarded as fashionable wore the plumed, beaver-felted hat. So great was the demand for beaver hats that a shortage of beaver developed in Europe. The insatiable demand for beaver lead to the discovery of beaver in North America and gave impetus to explore this new and promising area. The earlycomers to North America were not settlers but trappers looking for the beaver who were worth "their weight in gold." These trappers opened up the area for the settlers. The rest is history.

During this period the muff was another essential for the well-dressed person of either sex. An outgrowth of the fur-edged angel sleeve, muffs soon became large and elaborate, with various color combinations and alternating strips of fur in different directions. The "muff dog," a personal pet small enough to be kept warm in the owner's large muff, was a status symbol. By 1661, Saxon law limited ermine to high officials and aristocrats, while dyed sable was relegated to the professional.

Those who think men's and women's fur hats a recent fashion will find to their surprise that men and women of this period almost universally wore caps, tall hats, and berets of fur (the last adapted from the clerical berretta which became the university's mortarboard and finally emerged as the beret). Germanic women wore a huge *pelzkappe*, or large bonnet, usually in a long-haired fur. The hood retained its popularity and was made of velvet and lined or trimmed in fur. Military units, which began at this time to sport elaborate uniforms, often adopted a distinctive hat or cap of a particular fur as a regimental insignia. The bearskin *shako* and the *busby* of the Hussar are typical.

The short cape of earlier times remained fashionable, often copied in style by women but not worn in the same way. Greatcoats, cassocks and variations of fur-lined and fur-trimmed overcoats for winter wear grew from military garb, a major source of new styles in those days. The crafts and professions adopted many distinctive costumes. For example, the successful Italian doctor wore an ankle-length velvet gown with what would today be called a tuxedo fur collar, extending full length in front and forming a long center back panel.

BEGINNING OF AMERICAN FUR INDUSTRY (18TH CENTURY)

The forerunners of many world-wide fur companies started on the European continent in the 18th Century. Many of the furs used at this time are still used today. The hood and muff persist in popularity, as does the use of furs for linings, trimmings, caps, and gloves. Many of Watteau and De Lille's immortal portraits feature a muff of white lambskin and a fur wrap trimmed with marten over the typical pleated gown edged with fine fur. The bouffant hoop gown or skirt necessitated a short mantilla edged with fur, or a hip length cape-like pelisse lined and trimmed with fur. The shoulder cape pelerine, forerunner of the stole, was also used over these full garments. For more formal occasions, a cover-all hooded wrap bordered with fine fur was a fashion necessity. The Empress Josephine's high-waisted Empire coronation gown, which closed the century and set off a style trend of its own, was topped by a robe with an enormous train entirely lined and bordered with fur.

The 18th Century males loved fur. From Hungary came the carefully tailored close-fitting Hussar jacket, topped by a bearskin busby and covered with a decorative shawl or pelisse of leopard "carelessly" tied over one shoulder. Perhaps at no time in history was a stylish Frenchman's court costume more costly. It had to be lined with expensive fur. If he could not afford to buy it, the equivalent of over one thousand dollars enabled the would-be courtier to rent one for a day. The vogue for fur spread to the male bedroom, where the great men of the times wore dressing gowns lined with fur, a fashion that became almost universal in the American colonies. Common to all these garments, (coat, gown, or pelisse), was the *Brandenberg* corded fastening, which had originally been used on the Hungarian Hussar's jacket and spread over the fashion world. Periodically, as in 1973, this elaborate fastening now known as a *frog*, reappears in fur fashion.

THE AGE OF ASTRAKAN THE 19TH CENTURY

Examination of the styles of the 19th Century can be reinforced by perusal of magazines, newspapers, and periodicals which reported in whole or in part on the styles of the period. The introduction of photography made style communication much easier. The other permanent and distinguishing feature of the era was the emergence of *astrakan*, which we call *persian lamb* in the United States and which is known as *karakul* in Europe, Asia, and Africa. Furs mentioned previously remained in use, with fashion fads for "new" furs such as plucked sealskin, buffalo (for carriage robes) and monkey.

13

The "Empire look," which started at the close of the previous century, persisted through the Napoleonic period. A covering for a gown might be short, hip-length or full-length, but it was trimmed or lined with fur. The pellerine remained popular as an over-the-shoulder fur piece, but it was challenged by a long, snakelike string of fur called a *boa*, which was wound around the neck one or two times and let fall, often as long as the dress. The muff shrank to a diminutive size of little use as a hand warmer.

Men customarily wore fur-lined, smooth, cloth coats in black or brown, with fine collars of beaver and otter. This fashion persisted well into the 20th Century. It was not uncommon for the wealthy owner of a horse and carriage to dress his driver in such a coat. The beaver felt hat was a fashion casualty in this century, giving way to fabric hats.

The second half of this era witnessed a revival of voluminous mantles and cloaks, each style bearing some geographical or prominent person's name. Most were lined with the customary furs and elaborately tassled or fastened. Towards the end of the century improved fur-processing techniques gave impetus to the first

Fit for an empress: Only someone of Napoleon's rank could have provided this ermine-edged garb for his Josephine.

At the French Court of about 1830, these rich ladies added fur trims to skirts and sleeves to indicate social status.

attempts to use fur on the outside, rather than as a lining. Doucet of Paris is said to have been the first couturier to make fur garments with the texture on the outside.

The first persian lamb jackets were shown in Paris in 1885, followed by coats of seal and persian lamb in Leipzig in 1893.[2] Reproductions of fine ladies' gowns of this era almost invariably include fur in some mode. Examples include a trimmed shoulder capelet, pelerine, and off-shoulder dress trim.

FUR ON THE OUTSIDE—
THE 20TH CENTURY

It was not until the 20th Century that hair-out, sometimes called fur-on-the-outside, fur coats became an accepted fashion in the western world. The inhabitants of frigid countries like Russia, Siberia, and the Arctic regions had used this technique since time immemorial, as did prehistoric and "barbaric" peoples. These peoples used this technique not so much for the beauty of displaying the fur as for its superior heat-holding ability. However, it was not until improved processing had made the pelts manageable that civilized societies began to handle fur as a garment material in itself rather than as a lining, trimming, or accessory to cloth.

The history of fur styles is well documented and illustrated in many publications devoted to that subject. Magazine reports indicate that the only fur-out garments made before 1900 were short jackets of persian lamb and Alaska seal.[3] Other uses of fur in that period included hats, muffs, and carriage robes. By 1880, with the bustle on its way out, the length was extended to three-quarters. The full-length fur coat made its appearance with the onset of the 20th Century. Fashion plates of 1904 show fur greatcoats for both men and women designed to be worn in the newly invented automobile. From then on, succeeding fashion illustrations

Fur trimmings and accessories were important to this 19th Century lady, hands in muff.

reflect accelerated style changes, some bearing striking resemblance to the new style "swings" of 1947 and 1973. They varied in length and silhouette then, just as they do now, except that early styles were perhaps more elaborate than their modern re-creations. Long or short, straight or flared, horizontal or vertical, all can be seen in the styles of those years. A cursory perusal of fashion articles of those years reinforces the saying, "There is nothing new under the sun."

A major factor in the popularization of furs was the introduction just before World War I of the natural, uncorseted look by the French designer, Poiret. This trend has begun to be fully realized in the 1970 "no-bra" fashions. The bulky, protective fur coat of the pioneer open touring-car era was adopted by the young collegians of the 1920's. The long-haired, square-shouldered silver fox cape, emblematic of status in the '30's, appears in almost every night club movie scene made in those days. Men wore fine evening broadcloth coats lined and collared with persian lamb before World War I, but this old standby became another war casualty. Muff and scarf sets, cocoon-shaped wraps or mantles with fur outside, and brocades with fur trim were popular among the fashion-

Big, bulky furs for men became a fad during the 1920's. This group sports popular long-hair raccoon.

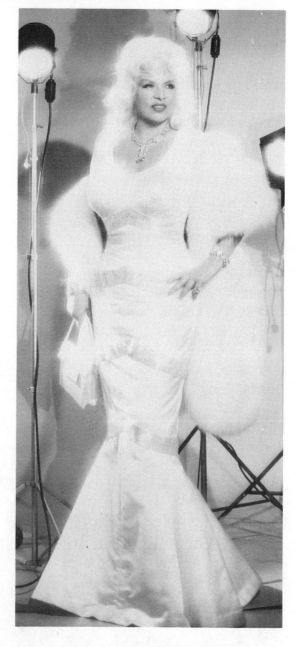

Two of the stellar movie stars of all time wear furs as a badge of glamour in these studio shots. At left, Jean Harlow is smothered in wild mink; above Mae West in satin and fox.

Varied tastes in fur for this
famed film trio: Marlene Dietrich,
upper left, dripping in ermine;
Barbara Stanwyck, left, in
dark mink; Betty Grable, above,
with full-furred fox.

Three who chose small pieces:
Ava Gardner, upper left,
in mink; Marilyn Monroe, left,
in white fox; Rita Hayworth,
coolly dangles ermine.

able women of the day. It was not until well into the '20's that more casual furs and styles to match came back in vogue. Nutria, civet cat, fox and barunduki were shown in little, informal items such as glove and shoe trim, dress and hat edging, and fur novelties. Early in the '40's, the pellerine reappeared, now called a *stole*, and carved a niche in the fur fashion picture that it has not relinquished to this day.

THE YOUTH MOVEMENT

Along with the population explosion, the 1960's may also be remembered as the decade when the wearing of furs bridged the generation gap and spread all over the world. New textures and transformations of the original forms of traditional furs rather than revolutionary new styles marked this decade. New combinations of skins, improved processing of existing skins never before used for garments, and assembled (pieced) furs in unique color combinations and patterns appeared. Furs began to interest the younger generation, both male and female. The age of the "fun fur," a description from which a Victorian fur-wearing lady would have recoiled in horror, arrived. For the first time in fur-fashion history, all age groups were interested in wearing furs.

The variety of wild furs, for reasons examined in depth later, has steadily decreased. To fill the gap and answer the demand for variety, the few remaining wild furs were treated and manufactured in new and excitingly different ways. For example, the common North American raccoon, traditionally worked in its natural long-haired form, was sheared to reveal a natural brownish pile-look, much like sheared beaver, which opened up an entirely new market for its use. Not content with this, innovators bleached the skins to give them another entirely different look more like fitch or small fox.

More of the domesticated animals were processed so that they could be used as fur. "Lamb" in furdom had traditionally meant either persian lamb or a sheared mouton lamb. Suddenly, more than a dozen new fur lambs, from Spain, Africa, South America, and China, were dressed and used. "Borrego," "kalgan," "icelandic" and "mongolian" lambs appeared on the fur scene as garments. In addition, garments from skins of horses, calves and other domesticated animals were shown for the first time.

Many of the remaining wild furs, some hitherto regarded as too heavy or too exotic for human wear in civilized society, also became popular. A sudden fashion demand for leopard spread to exotically patterned furs such as zebra, giraffe, jaguar, tiger, and lion, and returned to favor such neglected furs such as cheetah and ocelot. Eventually this trend became so strong that the continued existence of some of these species was endangered, giving impetus to a conservation movement discussed in a later chapter.

For the past five decades, the younger generation regarded a fur garment as a mark of middle or advanced age. A young girl from a well-to-do family might sport a coat, hat and muff outfit in rabbit or

In the 1960's, a big move to so-called "young," less costly furs: here, a bassarisk coat.

The young designers of the late '50's and early '60's who revolutionized and conquered the rest of the apparel field, turned their attention to furs. Encouraged by a few progressive manufacturers tired of spelling fur "m-i-n-k," they produced new effects, silhouettes and techniques, and modernized old procedures to create a new youthful image for furs. Undaunted by fur traditionalists who complained that "it had never been done that way," a group of designers, manufacturers and retailers risked reputation and capital to promote their new ideas. They were labelled "odd-ball," "way-out" and "sport" by their detractors. It took the first half of the 1960's and some casualties before these new concepts took hold.

By the opening of the '70's, fur was "young." Older women and many men were buying and wearing the styles, colors, and furs now identified with youth. The more radical youngsters, perhaps in another form of protest, resurrected and wore discarded fur garments, 20, 30 or more years old. For the foreseeable future, the widest possible acceptance of furs by the entire age and fashion spectrum seems fairly well established.

squirrel belly, but that was as far as it went for those in their teens to thirties. The fascinating film star, or socially prominent young woman of that age group might appear at the opening of the opera or a new movie in a glamorous white wrap of ermine or fox, but these appearances were rare and strictly formal.

FOOTNOTES—CHAPTER 2

1 Dr. Eva Neinholdt, *Hermalin* (Leipzig, 1964, Vol 4) p. 2.
2 Francis Weiss, *op. cit.*
3 R. Latto, "Alice in Fashionland" (*American Furrier*—put out by Sol Vogel, New York), July, 1948 p. 11.

3

The Fur Pelt: Processing, Labeling, Selling

3. The Fur Pelt: Processing, Labeling, Selling

WHAT IS A FUR?

When talking about the skins of animals used to make garments, the layman employs the terms "pelt," "skin," and "hide" interchangeably. However, the correct usage depends upon the exact nature of the fur being handled, its destined use, and the specific phase of the leather or fur industry involved.

A careful examination of a dressed fur skin which has been cut down the middle of the underside from head to tail, discloses several important things. The leather, which should be thin, soft and elastic, has from one to several layers. On most furs there is a dense, dull, curly mass of fine fiber, much like wool fiber. This warming protective layer next to the leather is especially dense on animals that live in very cold climates. On a fur seal, for example, there are approximately 300,000 of these fibers to the square inch.[1] This mass of fiber is most dense along the center of the back and becomes more sparse farther away from the back. Growing through this mass are the longer, straight, shiny, and usually darker guard hairs. The guard hairs protect the animal by keeping both the fur fiber and the leather dry and act much like a raincoat. Profuse and naturally oily the guard hairs can shed much water.

The term "skinning" originally meant the action of removing the skin from the carcass of the animal. If what came off was a combination of hair and fiber it was called a *pelt*. If it had little or no fiber and was composed almost completely of leather and hair it was technically called a *hide*. Modern style and tanning improvements use many hides with the hair left to make garments, breaking down the formal distinction between pelts and hides.

The leather industry, closely allied with furs, uses the same terms with slightly different meanings. To them, a hide is the outer covering of a fully grown animal of the larger species such as cattle, horses, camels, and elephants. Leather tanned with its original fiber and hair intact is called fur. Leather craftsmen, as opposed to those in the fur trade, regard a pelt as something tanned with the hair or wool removed.

Leather processors use the term skin in several ways. Generally it is the outer covering of an animal. More specifically, it is the outer covering of a smaller animal, or of an immature specimen of the larger animals, or the skin of a fur-bearing animal dressed and finished with the hair on.

Fur craftsmen often differentiate between long-haired and flat furs. However, the recent expansion of the shearing techniques with furs formerly used exclusively in their long-haired states has clouded this distinction.

SOURCES OF FUR PELTS
AND PROCESSING MATERIALS

Primitive man was a hunter of animals for food and clothing. After the animal was skinned and eaten, the female tanned the skin so that it could be used for clothing. The first tanning process may have been chewing the skin of the animal, a practice persisting today among the women of some remote Eskimo tribes. The saliva contains certain chemicals which act as tanning agents and the chewing process makes the leather soft and pliable.

When primitive man became a herdsman of lambs and other animals, their pelts were used for adornment. This practice is described in the early books of the Old Testament and is represented in prehistoric cave drawings.

All types of wild animals have been caged for sport or display for thousands of years. However the idea of confining live pairs of captured wild animals in a prescribed area and breeding them for their pelts is about 100 years old. This recently developed science of fur ranching has advanced and spread numerically and geographically. Mink, some types of foxes, chinchilla, nutria, and beaver are among the pelts produced commercially by ranching. As the need arises, attempts will surely be made to raise other species commercially.

The tightly controlled ranch, with pens for individual animals, permits the rancher to control breeding to improve the stock, to minimize illness and damage, and to experiment with various diets. The greatest gains from controlled breeding are improvements in color and texture. The mink breeders, principally those in the United States, have evolved over 200 hues and grades of mink, ranging from a pure white to an almost perfect black first shown at the end of the 1960's. The first commercial mink mutation was the platinum or silver blue which appeared in 1931. Tremendous progress has occurred in mutation mink breeding in the subsequent four decades.

Without becoming too technical, something should be said about how these new color phases are developed. Sometimes they result from careful mating of the best available male and female possessing the desired characteristics such as color, sheen, or size. Or a pup may appear in a given litter whose color, size, or sheen is markedly different from parents and siblings. In the early days of mink ranching such pups were regarded as freaks and considered a total loss. Early biologists called them Wasserman Sports, an educated way of dismissing a phenomenon without knowing why it happened.[2] Increased knowledge of genetics has given us greater insight into the biological changes which take place. New and better strains and colors of fur can thus be developed.

The breeding of fur animals has become international. The sale of breeding

Mink breeding has led to the development of many new colors and shades. A range is shown here.

stock is an important source of income for ranchers. For example, Scandinavia has developed a multi-million-dollar mink crop based primarily on the purchase of mink trios (a male and two females) from leading American mink ranchers. Occasionally, would-be ranchers in one country steal good breeding stock from ranchers in another country. A specific instance was the smuggling of a few pair of chinchilla from Peru to America in 1924 by an American mining engineer named Chapman.

Only since the 1950's has live American mink breeding stock been exported. Denmark, Sweden, Norway, Israel, Japan, and Holland are among the countries that have benefitted from this American resource. The Scandinavian countries as a group are challenging U.S. mink ranches in total production of skins. In 1971 they produced 11 million out of a total world production of 21¼ million.[3] Soviet buyers are purchasing Canadian breeders to improve their stock.

The persian lamb, which originated in Persia (now Iran) has been bred in much of eastern Europe and nearby Asia, including Romania and Afghanistan. Ironically, an animal similar to the persian lamb may well have been bred originally in the Russian state of Bukhara. There is evidence that more than 200 years ago Russian herders favored this animal, noted for its fat tail, as a food delicacy. The value of the pelt was considered a dividend.

In 1907 the first karakul lambs, seven ewes and two rams, were imported from Bukhara, (now Uzbekistan) into Southwest Africa. From these beginnings, involving some cross-breeding with local sheep, a worldwide commerce developed which produces several million pelts annually.[4] A story persists that the original blood lines began to disappear in successive generations and that by some sort of "under-the-table" deal Southwest African breeders were able to import additional rams after 1917 despite Russian restrictions against such exports. Whether this is true or not, the Southwest African "karakul" is now in contention for world leadership in its category. The pelt is lighter in weight, thinner in leather, and has a wavier and more moiré pattern than its competitors.

The important breeders of chinchilla, mink, nutria, persian lamb, or other ranched or herded fur in a given country usually find it to their advantage to form an organization to exchange information and regulate and promote their product. Among the more important are EMBA (Mutation Mink Breeder's Association); GLMA (Great Lakes Mink Association) in the United States; SAGA (Scandinavian Mink Ranchers), "Swakara" (South West African Karakul Association), and the Afghanistan Karakul Institute (Kabul, Afghanistan).

WILD ANIMAL PRESERVES

Hunting preserves, usually reserved for nobility, have their roots deep in antiquity. The idea of setting aside a given area of unspoiled land to preserve natural life in its original state has come to civilization almost too late. Our American national parks are a good example.

Another partially effective method of protection is to issue licenses which limit the killing of specified forms of wildlife. Both methods require effective enforcement. Effective enforcement is, however, not possible without the understanding and cooperation of the citizens in the areas. Unfortunately, many of the preserves are filled with poachers who are actively engaged in killing the very animals that the country is seeking to protect. These poachers have the support and protection of the local people. The profit motive is often the determining factor.

Sometimes the local farmers and ranchers will actively attempt to kill as many of the protected species as possible to rid the area of, what is to them, a pest and a hazard to their farms and flocks. The attempt to protect the wolves and eagles in the American west has faced this form of local resistance. How to bring all interested parties together to understand and support this need for preservation is difficult but necessary.

Perhaps the best example of the effective control and preservation of an endangered species is the Alaska Convention of 1911, which came after uncontrolled open sea killing had decimated the fur seal herd to near extinction.[5] Japan, Russia, Canada, and the United States agreed to a system of control, including supervised killing, to be managed by the American Department of the Interior. As a result, the herd has remained healthy and has increased from a low of 200,000 to the present level of about 1,400,000 which is close to the beneficial maximum for these animals.

The oldest ranched fur requiring herd control has been the sheep. Using sheepskin as fur has increased steadily with the advance of civilization because it is an easily obtainable and inexpensive source. For furriers, the outstanding example is the "karakul" (persian lamb) and its cousins. In 1967 a reliable estimate of world

The fur dealer gathers furs from all parts of world. These men are checking quality of such furs as mink, squirrel and muskrat.

karakul sheep production was approximately 31 million, half centered in the U.S.S.R.[6] About 40 other countries reported significant numbers. Generally, persian lamb ranching is profitable where there is large acreage of land suitable only for grazing and cheap plentiful labor. Present interest centers on the breeding of colors. White is now in the experimental stage. Grays, a long-time staple, are being developed in more shades than ever. Browns and black are also being bred, with the accent on lighter, flatter types.

WHAT'S THE NAME OF THAT FUR?

Nothing is seemingly simpler than to designate a fur by its name. However, this is not always true. Should the scientific name for the fur be used? If a fun fur is labelled *Lepus Coniculus* will we know that it is a rabbit? Will friends be impressed with the collar of a new coat if they are told that it's *Mustela Zibellina* instead of sable? Is *Castor Canadensis* more descriptive, impressive, or exact than beaver?

For everyday use it is more economical to use the colloquial or commonly accepted name for a product, whether it is a fur, a fabric, or artificial. Nylon and gasoline are clearly understood and preferable to their polysyllabic scientific designations. Most furs are similarly known by easily pronounced and readily recognized names. Sometimes there is a conflict over the commonly accepted name, such as karakul or persian lamb. Usually such differences are caused by linguistic variations.

In the first half of the 20th Century, segments of the fur industry began to imitate or copy each other by processing the more prestigious and expensive furs using a less expensive fur as a base. Often the substitute fur was a rabbit. Some gray color phases of this fur were especially bred or processed to look like chinchilla and were called chinchillette and chinchilla rabbit. Others were sheared, dyed, and processed and became beaverette, sealine, Mendoza beaver, and so forth. A major technical advance in the close shearing of the South American slink lamb developed a product which bore a marked resemblance to the Russian broadtail and was named American broadtail. This name was, of course, deceptive.

Perhaps the most famous of these substitutes was Hudson seal, a staple of the industry as far back as the early 1920's. It grew out of the demand for Alaska seal, then, as now, a prestigious and expensive fur. Certain grades of muskrats from the northern United States and neighboring Canada could be plucked, sheared, and dyed black, so that when properly worked, the resulting garment closely resembled black dyed Alaska fur seal. The garment was to be light, durable, and usually less than half the cost of the genuine seal. One of the industry's largest fur skin processors in the 1920's, A. Hollander & Sons, had a big plant in Newark, N.J. employing several hundred workers. Many large manufacturing firms specialized in this item. Hudson seal gradually lost favor even before the advent of labelling regulations, although embittered furriers ascribe its total demise to this latter blow.

Just as there is no natural American broadtail as such, there is no natural fur seal indigenous to either the Hudson River or Hudson's Bay. Hudson's Bay has some seals, but those are more likely to be the ringed seal, the bearded or square flipper seal, or the common ranger seal.

The concept of consumer protection by truthful labelling developed during the second half of the 20th Century. The Federal Trade Commission, acting as required by the Fur Products Labelling Act of 1951, promulgated a set of approximate rules and regulations for the fur industry. It included all types and kinds of furs and required that the labelling be in English. The use of an adjectival name which tends to mislead or give a false geographical origin was prohibited, with some exceptions. Trade or coined names describing a fictitious fur such as those in the examples above was prohibited, unless followed by the legally accepted description, for example, "American broadtail, processed lamb."

Under these regulations, the country of origin of a given fur must be clearly indicated in all labelling, invoicing or advertising and preceded by the phrase "fur origin" or "country of origin," for example, "dyed muskrat, fur origin Russia." If two different furs are used on one garment, each must be correctly labelled. During the 1950's other regulations were introduced covering dyeing and bleaching, the use of pieced furs, and the use and sale of used furs.

To enforce all of the above regulations, all fur garment manufactures must obtain an individually registered identification number which must appear on the sales ticket of each garment. The text of the act also lists all animals known to be used as fur at the time and the name and scientific designation of each, with provisions for the addition or subtraction of names as required.

The net result of these regulations is to insure that the consumer is fully aware of the exact nature of each fur item. Legitimate furriers, in common with the consumer, hail these regulations as protection against the rare, unscrupulous seller. The furriers also are quick to point out that in some instances the deglamorization of the names of previously popular furs as a result of this act has adversely affected their sales. The major fur associations for the more important furs have worked diligently to overcome this deglamorization by trade-marking fanciful names for the various shades of fur they have bred. By far the most successful has been the American mink breeders associations. The best example is EMBA, whose trademark registered names include "Rovalia," "Tourmaline," "Morning Light," "Azurene" and "Jasmine."

The country of origin requirement serves the dual purpose of indicating origin for tax and embargo laws and to prevent substitution of apparently inferior skins.

The cape seal from the Cape of Good Hope is a fur seal which, when dyed black, bears a close resemblance to the Alaska fur seal. However, it is not quite as dense nor does it wear as well. In this instance "country of origin" is important. The same holds true for Siberian and Canadian squirrel. The former is fuller, denser and consequently more valuable than the latter.

Unfortunately, the desirability of certain pelts from a given section of a country, or for that matter, the country itself, tempts unscrupulous handlers to ship pelts caught in a less desirable area to the main shipping center of the more desirable section or country so they can be sold as originating there. Any experienced auction house, assorter or skin dealer who has spent a lifetime handling his specialty can tell at a glance when it has been mislabelled. Recently this evasive technique has been used to circumvent strict conservation regulations. The poacher illegally transports his skins to a neighboring state whose laws are less stringent or whose custom officials are more accommodating and ships them out legally. On the continent of Africa, with its multitude of independent countries many of whom have similar animal ecologies there are hints of the existence of this practice.

For ranched products like mink or fox, the ability to detect the country of origin by inspection has been all but obliterated, especially in the medium grades. In many instances all stem from the same breeding sources, constantly replenished by the exchange of additional live breeding stock, so that distinct national characteristics become less apparent.

On balance, however, both the consumer and the legitimate furrier, whether on the manufacturing or retail level, are offered protection from unscrupulous competitors. For these reasons they both favor such regulations even though they complicate record keeping and billing.

LABELING AND BILLING

The practical result of the foregoing regulations is exemplified in the typical tag and bill of sale.

A properly executed bill of sale should contain the minimum information as indicated on page 33. If paid on delivery the customer may ask that receipt of payment be indicated on the sales check. The seller will gladly oblige, usually by hand-written notation or with a "paid" stamp, indicating number and date of check and manner of payment.

PROCESSING THE SKIN

Our American trapper still prepares his pelts in much the same way as did his predecessors. He skins the carcass, sometimes by removing the skin like a pullover sweater without opening the belly and sometimes by splitting the skin down the belly and paws then removing it flat. He then scrapes off the flesh down to the leather and stretches it on a form or flat and allows it to cure or dry in the sun. The cured pelt is salted to preserve it, and shipped. This technique has been used for many centuries.

Even though the trapper's catch is skinned and stored under optimum conditions, it will deteriorate in both hair and leather quality if held too long and not tanned (dressed) within a reasonable period. The leather develops weak spots called burns and the follicles of fur hair and fiber tend to become dull or fall out

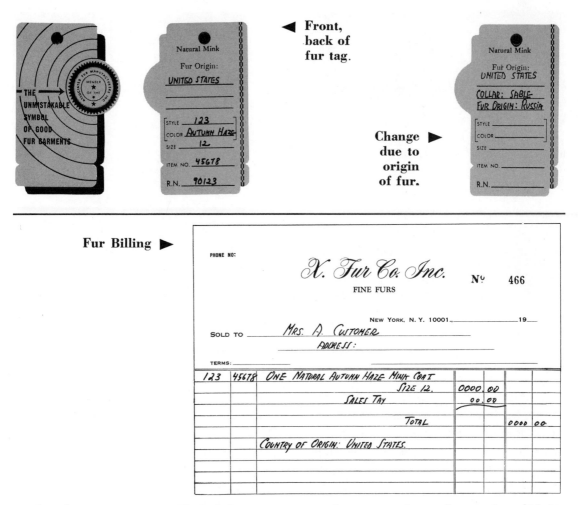

◄ Front, back of fur tag.

THE UNMISTAKABLE SYMBOL OF GOOD FUR GARMENTS

Natural Mink
Fur Origin:
UNITED STATES

STYLE 123
COLOR AUTUMN HAZE
SIZE 12
ITEM NO. 45678
R.N. 90123

Change ► due to origin of fur.

Natural Mink
Fur Origin:
UNITED STATES
COLLAR: SABLE
FUR ORIGIN: RUSSIA
STYLE
COLOR
SIZE
ITEM NO.
R.N.

Fur Billing ►

PHONE NO:

X. Fur Co. Inc.   Nº 466
FINE FURS

NEW YORK, N. Y. 10001,_____19__

SOLD TO _____ MRS. A. CUSTOMER _____
ADDRESS:

TERMS: _____

123	45678	ONE NATURAL AUTUMN HAZE MINK COAT				
		SIZE 12.	0000	00		
		SALES TAX	00	00		
		TOTAL			0000	00
		COUNTRY OF ORIGIN: UNITED STATES.				

in the dressing process. This deficiency often appears in garments made from skins in localities where no adequate tanning facilities exist or where the shipping of skins is irregular. The apparently low price paid for such garments by tourists in certain countries in which the animal is indigenous, has often turned out to be a poor bargain. The skins soon become stiff, crackly, and dried out. Worst of all, they cannot be re-treated to improve or restore their condition.

Skins, both dressed and raw, are shipped to the auction houses. These auc-tion houses are located in such world fur centers as New York, Winnipeg, and London.

FUR PELT BUYING

The individual who buys the skin is usually a principal of the firm with experience in buying and financing the purchases. Sometimes the volume requires the services of a full-time or part-time skin buyer specialist. The latter is more common, with the buyer dividing his serv-

Raw mink skins ready for auction. Here you see the leather side of pelts, assorted by color, size and quality.

ices among two, three, or four firms, for whom he buys at a percentage fee. His know-how must include not only a thorough knowledge of raw and dressed skins, but must also encompass an up-to-the-minute grasp of the current value of skins. He must also be completely familiar with each firm's needs, policies, and business goals.

AUCTION PROCEDURE

The potential buyer is notified of the auction dates by advance notice, trade-paper advertisement, and catalogue. He is allowed ample time before the auction date to examine the lots to be auctioned. Preliminary examination enables him to determine which lots meet his needs and what they are worth to him. If the skins are raw, estimates of potential value become somewhat more of a gamble.

There is a constant struggle between the owner and auction house on one side, and the buyers on the other, as to whether the skins should be sold raw or dressed. When the fur is of a type that will not deteriorate in the raw state, the former would prefer not to have to undergo the cost of dressing the skin, while the latter would like to be certain of the quality before buying. The buyer of raw skins will be forced to rely, to a great extent, on his previous experiences with the skins purchased from a particular fur farm or rancher.

The New York auctions are conducted, as in many other countries, by secret as opposed to open bidding. The auctioneer on the podium, with the assistance of spotters on the floor, receives bids by a prearranged private signal from the buyer. The buyer makes good his bid by a deposit before leaving, with the balance due within a specified time. In some instances, parts of the purchase may be left with the auction house to be "cleared out" by payment of the balance, plus a fee for storage and financing. The auction house is paid a percentage of the sales proceeds, usually 5%, by the seller. Lots can be put up by the owner with a stipulation that they can be withdrawn if they do not bring a minimum price.

The English and American auction houses are privately owned and managed. On the other hand, Scandinavian auction houses are often owned by breeders' associations. To promote sales, these breeders permit intermingling of lots (intersort-ing) which promotes uniformity of strings sold, a feature which is most attractive to the potential buyers. When the breeders are in charge, they have the power to set a floor or bottom price under which they will not sell, a power they invoked in the early '70's. American ranchers have fought this development so far, preferring to rely on individual reputation and advertising. This typically American propensity for promotion of individual excellence hopefully will not have to give way to a more economical method of marketing such as co-operatives.

Another much desired development is the reduction of the number and variety of these auctions. This has already been accomplished with American and Canadian seal auctions. It simplifies fur auction sales attendance. The apparent

Fur Auction in action: Men seated here will have examined furs prior to sale. Auctioneers, extreme right, are recording bids.

minute, modern, temperature and humidity controlled building with a suspended roof to give the buyers an unobstructed view of the sale. The contrast is striking between the confidence shown in the future of the fur skin business by the Scandinavians with their co-operative auction house-marketing and the decline of commissioned privately owned auction house-selling in the United States.

DRESSING

Whether before or after the auction, the skins must be dressed or tanned. Modern fine fur dressing is an art mastered in only a few countries where the volume, financing and, above all, the necessary skills are available. Like everything else in furs, this knowledge has diffused from one country to another.

The earliest known leather tanning center was probably in Morocco. Prior to World War I, the Austrian city of Leipzig was the world center for the dressing and dyeing of persian lambs and other furs. The center shifted to the United States in the 1920's. At present, West Germany, Italy, England, France, and the U.S. are the most successful, with the U.S. holding the lead for most furs.

Available records suggest that animal urine was one of the first leather softeners and anti-putrifants used, a practice that lasted well into the 19th Century. Other

redundancy of fur auction establishments in the New York market has already brought on the closing of one of the major auction houses in late 1970.

The American mink ranchers show signs of threatening to by-pass the auction technique altogether. They favor selling by commission agents directly to the manufacturers, thus also downgrading the dealers. The commission agents act for the ranchers or their co-operatives. A start has apparently been made in this direction by the formation of a marketing outlet for the Northwest fur breeders co-operative.

In contrast, the Oslo Auction Conglomerate, dedicated with much pomp and ceremony in 1968, is an up-to-the-

Dressing and Dyeing Pelts: These large vats contain chemicals that transform raw skins into beautiful, lustrous furs.

ancient tanning solutions included the bark of certain trees, including hemlock.

The tanning process changes the stiff, heavy, fleshy leather and dirty, oil-matted hair into a malleable, stretchable, light, thin, soft, creamy white leather as fine as that of any expensive glove. The hair and fiber are separated, cleaned and fluffed up to their maximum beauty. Most important of all, chemical treatment fixes the hair and fiber in their follicles and permanently protects all parts against the natural bacterial action which normally breaks down and destroys all dead animal matter.

The original natural oil of the leather is replaced by another oil which must be worked back into the dressed leather to preserve its life, softness, and durability just as shoe polishes help preserve shoe leather. The cost of such processing ranges from a few cents for rabbits to several dollars each for the larger or more expensive pelts. New developments in fur dressing include an enzyme system which, it is claimed, virtually eliminates most of the mechanical fleshing (scraping) and drastically cuts processing time.

The technological advances in fur dressing have, in general, aimed to reduce the weight and thickness of the pelt, in keeping with the needs and desires of the 20th Century customer. This contrasts with the traditional desire for a fur which was thick skinned, bulky, and heavy. Travelling from centrally heated homes in a heated conveyance to a centrally heated hall, restaurant, or theater, the modern man, or woman, of fashion does not have to maintain body warmth by very warm furs over several layers of cloth, as did his grandparents.

The demand for lighter, more fashionable furs has adversely affected the popularity of such furs as the comparatively heavy Russian persian lamb and the western U.S. grades of beavers known as "blankets." The processors are motivated to improve the dressing process to produce a thinner, lighter leather, because the leather is the significant factor in the weight of a pelt. There are practical limits to such improvements. Shaving the leather too thin may create a weak spot, or weaken the follicles in which the hairs or fibers are anchored.

ALTERING THE APPEARANCE OF THE FUR PELT

One of the characteristics of modern furdom is the wide variety of techniques that alter the appearance of fur to something original and different, or process fur so that it resembles some other more expen-

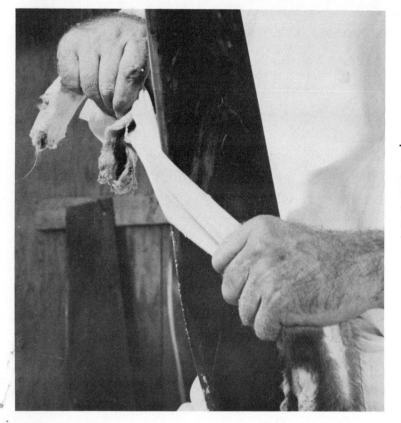

"Flesher" scrapes fur pelt on fleshy side, taking off excess materials, and reducing skin to desirable thickness, pliability.

CHART OF COMMON FUR-COLORING TECHNIQUES

Name	Method	Purpose	Distinguishing Feature
DYEING	Total immersion of fur into a mordant and then into a dye.	Change color of entire unit, hair, leather and fiber.	Leather dyed to same or similar shade as fur hair and fiber. Hair sometimes singed from over-dyeing.
BLENDING	Adding color to tip of hair or fiber by feather brush or air brush.	To color only the tip of the guard hair or fiber on all or part of skin as needed.	Color addition on tips can be detected only by close microscopic examination or by experienced furrier.
PRINTING	Application of color in some pattern (usually by silk screen) to prepare hair or fiber or pelt of plates (pelt or pieces).	To imitate more expensive product. To create new design effects on fur, creating variety.	Color of pattern penetrates partly on hair or fiber length. Pattern crosses seams.
BLEACHING	Application of bleaching agent to part or all of hair and fiber.	To lighten part or all of color of skin so as to present different appearance. To whiten all or part of skin as needed.	Leather often made weaker and powdery, unnaturally white. Stitching may be weakened. Must be compared with natural or untouched skin to detect difference.

sive fur. These techniques include dyeing, printing, blending (tinting), plucking, shearing, grooving, and tipping. There have been some attempts to paint fur as canvas. So far this is a novelty.

It is possible to color furs in several ways. In simplified form, they are shown on the above chart.

Available historical records about the dyeing of furs indicate that the earliest known preference was for red, a fur color fashionable over 4000 years ago.[7] Furs which were naturally red were most desired. Others were valued for their ability to respond to available red dyestuffs. The excesses of hidden blending, dyeing and "touch-up" of a pelt to improve its poor, natural appearance, or to make it resemble more expensive species were common enough by the Middle Ages, to cause historical comment and require law-making, much as today.

NATURAL OR "COLOR ALTERED"

Technical developments in the art of fur dressing in recent years include the addition of iron and copper salts in very small quantities to the basic water soaking which is part of the process. The dressers contend that this maintains and restores the original color of mink and some other furs. The Federal Trade Commission, which has jurisdiction, ruled after much hesitation and indecision that 300 parts or more per million parts of water measured by volume in the dressed leather constituted addition of color and required classification as a "color added" rather than a naturally dressed pelt. The distinguishing characteristic of a pelt heavily dressed in this manner is a purplish tinge to the leather of the pelt.

The January 7, 1969 FTC order spelling out the specifics of this ruling and requiring the identification of the nature of the dressing of a given skin (above or below the limits of addition of copper and iron salts) by imprinting on the leather black, yellow, or white marks, brought a stormy protest from every segment of the industry. The technical nature of the strong arguments presented are not germane to this book, but they were sufficient to cause the Federal Trade Commission to delay enforcing its edict. Presently, skins dressed "naturally" in conformance with the limits set down by the Commission are stamped "natural" in black ink. Skins not bearing this stamp are presumed to have been dressed with some form of color addition beyond the F.T.C. regulations.

A letter from the attorney for the Fur Dressers Bureau of America dated July 13, 1970 states in part; "The enforcement of certain portions of the amendment (F.T.C. order of January 17, 1969) is in a state of limbo because, subsequent to the order, the Commission was deluged with complaints from nearly every segment of the industry."[8]

All of which proves that government "experts" are not always omnipotent and that such experts and agencies sometimes respond to reasonable objections. This is how matters stood until April 5, 1971, when the Federal Trade Commission finally ruled that the pelts could be stamped "dyed," "color altered," "color added," or "natural."[9]

PLUCKING AND SHEARING

Early in the history of furs it was discovered that amphibious or aquatic furs like beaver, seal, otter, nutria, and muskrat could be changed in appearance by plucking out the long, straight, shiny guard hairs and thus exposing the dense, soft, dull, wavy fiber underneath. This fiber could be sheared to a shorter and predetermined length.

This technique was applied to American muskrat, which was then dyed black to create "Hudson Seal," one of the most popular furs during the first quarter of the 20th Century. The advent of the F.T.C. Correct Labelling Regulations drove this mythically named fur off the market. The original historical demand for beaver skins which played such an important part in the early development of the North American continent, was actually a desire for the fiber of the beaver, which was sheared completely off the skin to make one of the finest felts obtainable.

At first, shearing was limited to machine cutting of the fiber to an even height. Modern developments include contour shearing, in which the height of the shearing is gradually sloped to give a contoured look. Natural sheared beaver is a good example. Another variation is to cut several sharp parallel grooves into the fiber in narrow patterns. The first use of this grooving technique was on sheared white rabbit to imitate ermine, using grooves about three inches apart.

FOOTNOTES—CHAPTER 3

1 Inside Story of the Fur Seal, The Fouke Co., Greenville, South Carolina, 1969, p. 4.
2 "Birth Defects in Animals," *Fur Age Weekly*, New York: March 1, 1971, p. 2.
3 O. Brager-Larsen, "1971–72 Mink Crop," *Fur Age Weekly*, New York: March 1, 1972, p. 2.
4 Jurgen Thorer, "The Future of the Persian Lamb," *International Fur Review*, London, March, 1968, p. 18.
5 "Fur Seal's of Alaska's Pribilof Islands," Fish and Wildlife Service, Washington, D.C.: U.S. Dept. of Interior, p. 1.
6 Ivy Sharpe, *op. cit.*
7 Terence Tuttle, "How to Grade Furs," Ottowa, Canada, Dept. of Agriculture; No. 1362, 1968, p. 13.
8 Stringari, Fritz, Fiott, and Burwell Petition to Federal Trade Commission on Behalf of Fur Dressers Bureau of America, Chicago, May 12, 1969.
9 "FTC Amends Pelt Marking Requirement," *Fur Age Weekly*, New York, April 5, 1971, p. 2.

4

The U.S. Fur Industry: What Makes It Tick

4. The U.S. Fur Industry: What Makes It Tick

THE FURRIER'S CALENDAR

As in any seasonal industry, the furrier's calendar is peculiar to itself. It begins in late February or early March. The organization, previously running at low gear finishing off the remnants of the previous year's orders, begins to shift its attention to the coming season. Information and ideas about style trends have been collected and form part of the basis upon which the style "line" for the forthcoming season will be built. Of course, some commitments for skins have already been made, especially if they require several weeks for dressing, dyeing, or other processing.

Another factor affecting the furrier's timetable is the requirement that pelts be harvested when their coats are prime, that is, when the fur is of maximum value. This is usually soon after the first frost. The animals are large and full from the warmer weather and the onset of colder weather tends to make their fur denser, fuller, and more lustrous. The owner, whether he be trapper, rancher or skin dealer, naturally wants to sell his fur skins as quickly as possible to avoid spoilage. He also wants to turn his harvest into cash. Other factors that will play a part in determining the furrier's calendar include his previous season's selling experience, new fashion trends, and changes in the economic picture. Manufacturers and re-

tail stores have recently been "pushing the seasons," by presenting the new seasons' wares earlier and earlier.

Only a small percentage of furriers, manufacturers, or retailers can afford an innovative creative designer on their staffs. Many have a skilled craftsman with the title of designer who is actually more of a pattern adjuster, grader, fitter, and alteration hand than a true designer. For most firms, these abilities are more useful and valuable. Most new ideas tend to come from many outside sources.

A major supplier of new ideas is the *public designer* of fur styles. Two or three dozen of these firms, each with a small staff, work up their ideas in *canvasses* (*toiles* in Europe) each season and show their creations to interested buyers. Importers and salespeople for European design houses also schedule seasonal showings at an advertised location, or come to the showroom of a known potential buyer. Scheduled showings are also part of the presentation techniques used by the larger local public design houses.

The manufacturer, wholesale or retail, buys a paper pattern in the selected style and in the size or sizes that he needs. It is common for several manufacturers to share the cost of purchasing the patterns. They make several copies of the original, thereby cutting expenses. Often the pattern is retested in canvass and modified according to taste and need.

Since these decisions may make or

THE PUBLIC DESIGNER: MAJOR SUPPLIER OF NEW IDEAS ON FURS

(Left) designer sketches. Lower left, canvas is being prepared. Note canvases on hangers. Below, coat canvas has been completed.

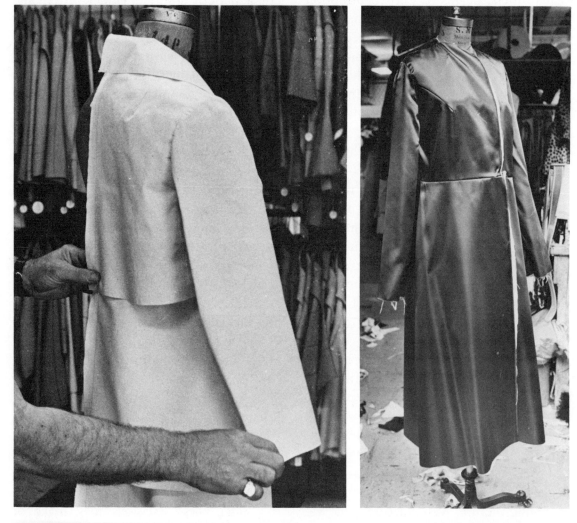

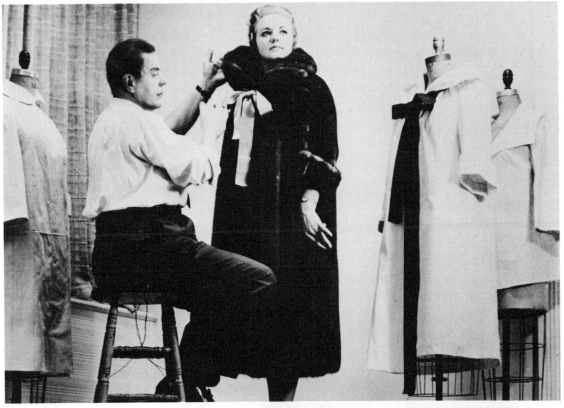

**Final fur fitting:
Designer, above, checks coat
against canvas. Right,
mink coat and canvas.**

ruin the season, they are almost always made by the principals of the concern, especially those principals responsible for sales. To round out the forthcoming line, the most successful of last year's styles are up-dated and included.

The smaller and less innovative firms have for many years lacked style consciousness. The same basic styles were shown year in and year out. Many believed that with their limited finances they could not "take a chance" on the new styles and ideas. Fortunately, many of these firms have now discovered that they cannot "take a chance" on giving the customer something which she already has rather than something which she really likes. Innovation has become widespread. Today, it is the new ideas and styles that spell the difference between success and failure.

A newer source of fur designs is the successful, well-known designer, from another part of the apparel industry, perhaps based in another country. He, or she, franchises his name and ideas to one firm or chain. Big name milliners, coat and dress designers, and even men's clothing creators who have had success and fame in their own field are used. The industry also has several fur designers who have attained world renown on their own.

From all these sources a sample line is made up. For the New York market, it is officially shown during a three-week period in early June known as "Fur Week." These showings are often elaborate and take place in prestigious surroundings. The West Germans show at the Frankfurt Fur Fair, usually held the first week in April. Paris and Milan have their own shows at about the same time. The initial reactions of buyers, who, as far as possible, visit or keep track of all the showings, determine the directions early production will take.

A trip to any major fur market is expensive and time consuming for any buyer except those located in or near it. A necessary adjunct to the success of a fur business which is too far for frequent buying trips is a permanent representative on the scene who knows the fur business, the establishment's policy, clientele, and objectives and will make purchases to fill these requirements. The use of the fur resident buyer in the United States is apparently the best answer to this problem. Permanently based in the fur district, he or she does the buying on orders on a fee basis, usually about 5%, servicing several non-competitive accounts to round out the business.

WORLD FUR MANUFACTURING CENTERS

The world fur centers have ebbed and flowed historically as the sources of furs and available craftsmen have changed. For nearly three quarters of a century almost half the fur garments made in the world have come from New York City. Ninety-four percent of the total dollar volume of fur shipments in the U.S. in the past twenty years has come from New York State,[1] which for all practical purposes means the New York City fur district.

There are, of course, many other

Birds-Eye View of New York Fur Market: Within the circle shown here, a high percentage of the world's furs are made.

world centers of wholesale fur garment manufacturing. These include Chicago, Montreal, Toronto, Frankfurt, London, Paris, Kastoria and Milan. Recently, Tel Aviv and Tokyo have developed significant manufacturing centers. Israel is planning a department of fur technology at the Shenker College of Fashion and Textile Technology. Canada has already begun such a school in Montreal. Attempts have been made by this editor to initiate such a course at a higher level in New York, but the progress, to date, has been slow.

Eurofur, one of the world's large fur garment plants, is located in Padua, Italy. The Birger-Christensen factory in Denmark is said to employ 200 workers. A multi-million-dollar plant is being planned for Hakodate, Hokkaido, Japan. It will start with dressing and move on to manufacturing.

There are reports that the U.S.S.R. has two plants which, in sheer size, surpass anything in the rest of the world. They are completely vertical. That is, in one plant they take in the raw skins, dress and process them, and make them into fur garments. One factory in Kazan is said to employ 9000. Another in Moscow employing 6000 is devoted to trimmings for hats and coats. Less and less of the U.S.S.R.'s production of raw furs will be available for export as more of its 250 million people dress in furs.[2] There is no present indication of any desire on the part of Soyuzpushnina, the fur division of the U.S.S.R. to sell ready-made garments in foreign markets, because of tariff barriers and domestic needs.

The movement of fur garment manufacturing away from traditional centers such as New York seems to be accelerating. The fur industry, like many other skilled and well-paid crafts in the United States, is feeling the competition from other countries with lower standards of living. The recent dollar devaluations have, however, given a tremendous boost to the export of American fur garments. Foreign customers find that the American product is not only superior in craftsmanship, but now has become in many cases cheaper than their own. There are some manufacturers who have all but given up manufacturing for the domestic market in favor of the foreign market.

Willingness by the West German industry to develop and use improvements in fur machinery has put the Germans ahead of Americans technologically. The apparent reluctance of the younger generation all over the world to enter a trade which takes so much time to master should not be overlooked. This is part of a general reluctance to work in a factory as opposed to working in an office or as a professional. The unbiased observer will, however, admit that American craftsmanship is still supreme. It is also encouraging to note that at the High School of Fashion Industries in New York City, which offers the only course in fur manufacturing in the United States, the number of students enrolled in the department has doubled and re-doubled in the last two years. Perhaps, the trend away from the skilled trades is reversing.

FINANCING AND MERCHANDISING

Most U.S. fur manufacturing firms are under-capitalized. It is not unusual for an active New York firm to do a yearly volume of business ten times its capitalization. This ratio, in a business where the cost of the raw material can fluctuate so radically, increases both the profit and loss ratios. To finance his business, the average furrier has recourse to limited credit supplied by specialist banks or to credit advanced by his suppliers. It is curious that, outside of the U.S., bankers regard a stock of garments as an asset, while in the U.S. they tend to assess it as a liability. To add to the banker's problems, some of his customers, forced to order and buy well in advance of the selling season, may pay for their purchases with long-term notes that are sometimes interest-bearing. This increases the cost of

FLOW CHART: FROM CATCH TO CONSUMER

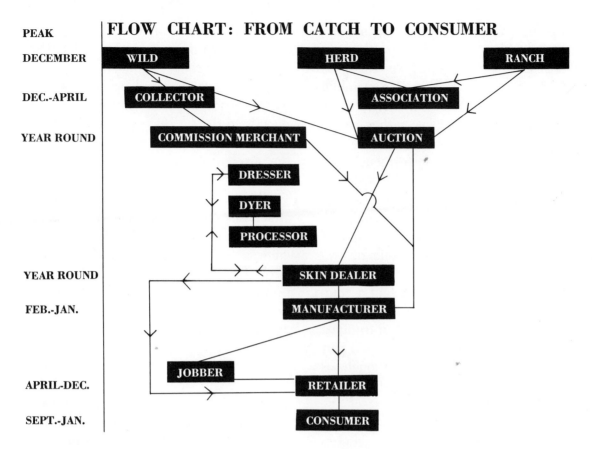

doing business all along the line.

The memorandum technique of sending garments to select customers on approval is no stranger to the furrier. Many attempts over the years have failed to eliminate what is at best a nuisance or at worst an expensive gamble. Sometimes a memorandum shipment is arranged for a limited time in return for a well-advertised sale. The memorandum technique has in a sense replaced the travelling fur salesman. Spiralling travel costs and the dangers of theft have reduced this once prominent sales technique to a vestige of its former importance.

The obvious advantages of investing money in the better known and more suc-

cessful fur manufacturers or retailers has prompted institutions or conglomerates to become partners in or owners of several top houses. Once absorbed, they have all the capital necessary to streamline and exploit their already well-established reputations. How far this will go is at present unknown, but it does open an entirely new horizon for the more prestigious firms in the fur business. With the financial backing and business know-how available this may lead to a new break-through in fur garment merchandising.

Traditionally, fur garments have been sold by specialty shops and fur departments of large department stores. In the past two decades, the all-fur specialty shops have declined in number and importance. Many fine establishments formerly exclusively devoted to fur have added other lines of apparel to maintain their volume. In New York City, many world-famous names have either closed up or moved and become the fur salon or department of a large department store.

The remaining well-established specialty shops and department stores have had to abandon, in large measure, the maintenance of their own fur garment manufacturing plants, with one or two notable exceptions. It is even more difficult and costly to get and retain a complete staff in this field than it is for custom dresses and suits. Fur-trimmed garments by and large are considered part of the coat department even though, as in fur-lined garments, the furred part represents more than half the cost of the item.

When a purchased garment requires substantial adjustments, it is often sent back to the manufacturer or to an independent manufacturer who takes on the repair work at a fee. All of this assumes, of course, that there is someone at the store who understands what is to be done and knows how to take the necessary measurements for that kind of alteration or repair. This is not always the case. Like many other departments, and perhaps more so, the fur department or fur specialty shop suffers from a lack of knowledgeable sales personnel. What customer can have confidence in a salesperson who doesn't even know the name of the fur garment unless he looks at the attached label?

The purchase of even a medium-priced fur garment is a major decision requiring personal confidence in the seller. Perhaps this explains the increasing tendency of many customers to go directly to the fur garment manufacturer. This saves money and gives the customer the assurance that she is dealing with someone who knows his product.

Retail sellers of fur have tried to attack this problem. They have discussed the possibility of starting a training program in fur merchandising in New York City. This course would train the salespeople of fur in those aspects of the business needed to make them knowledgeable about their product and helpful to the customer.

New York City, with the largest aggregation of fur manufacturing establishments, still has scores of specialty shops and department stores which compete successfully for the available business volume year after year. These successful operations possess a core of knowledgeable staff and management which make them click.

The new fur garment selling technique of the 1970's is the separate fur boutique. Originally set up in more progressive stores to cater to the youngsters who did not feel comfortable buying furs in the traditional fur department, this sales technique has mushroomed. They have expanded horizontally and vertically with new and unusual fur items. Many show not only apparel but also such items as rugs, throws, and bed covers. At first these boutiques stocked only budget items for the young, but they have gradually reached upward to the more expensive furs. Their success may come from their psychological appeal to the not-so-young customer who likes to feel still able to wear the clothes designed for the younger set.

WIDENING THE APPEAL OF FURS

Today there are many creative designers in the apparel field who have applied their talents to the fur industry. In the late 1920's, a mass importation from this source of designs proved to be a disaster. Not being adapted for fur, the garments proved to be misaligned and almost impossible to assemble.

The vogue for youthful furs, nurtured through the first half of the 1960's, was caused by innovation in the modern styling of formerly rarely used furs. Puma zebu (water buffalo), giraffe, lion, tiger, guanaco, horse, and zebra were among the furs that appeared in apparel form for the first time in Western Civilization. The designers were successful because they adapted themselves to the handling of fur as such instead of trying to treat it as cloth. Furthermore, the traditional furrier was now more than willing and ready to work with anyone who gave promise of widening the fur merchandising market so that it would grow as fast as the rest of

American industry. Having seen their craft virtually reduced to a one-fur industry and losing ground steadily in maintaining its share of the apparel business volume, they were in a frame of mind to co-operate with other segments of the industry.

The silhouette of fur garments coincides in general with that of the rest of the fashion industry at any given time. Among the smaller fur pieces there may be style developments with no counterpart in fabric, but they usually complement them. Sometimes it is the fur designer who originates a new variation of an old style, followed by his counterparts in ready-to-wear, but more often the reverse is true.

A good example is the 1947 fashion season when the Dior full-length garment (the New Look) burst like a bombshell on a fashion world accustomed to short dresses. The fur industry, which was in the midst of working up its 1947–48 fall and winter line, came to a floundering stop and marked time uncertainly trying to ascertain if this daring innovation would catch on or fizzle out. After a full season of tryouts in 1970, fashion history once again repeated itself with the maxi-coats. Would they take off or fizzle out? Ulcers can easily be the result of such uncertainty!

FOOTNOTES—CHAPTER 4

1 *The Fur Market*, Morton Research Corp., New York, 1970, Chart 3.
2 Bernard Leason, "Domestic Demands Pinch Soviets Fur Industry," *Women's Wear Daily*, New York, Aug. 18, 1970.

5

A Special Art:
Fur Design and Fit

5. A Special Art: Fur Design and Fit

INTRODUCTION

Our purpose is to suggest some of the problems inherent in designing fur garment styles. First, a finished fur garment has no give, unlike most fabrics and knits. The fur leather may have originally been resilient, but the finished garment has been permanently "set" or reinforced to hold its shape. The ease must be built-in, a factor which distresses many customers who, especially in the fitted styles, would like to have their fur garment as tight and clinging as a knit dress or sweater.

Despite the rapid progress of the fur designing industry, many furs are too heavy to be worn with comfort when assembled into a garment. The decline in popularity of Russian persian lamb may be traced in part to its comparatively heavy weight. Another reason for its decline was the change in style which decreed that persian lamb was a fur for the older and less style conscious woman.

The ingenuity of the designer can overcome this problem of weight in several ways. The fur can be used in narrow, small garments, interspersed with lighter material such as ribbon or leather, or it can be combined with other lighter

Ingenuity to a high degree is shown in this full-length blue fox coat, designed with skins horizontally.

weight furs or fabrics. A feature of the 1969 Frankfurt Fur Fair was the introduction of persian lamb in which the leather side of the skin was treated and embossed so that it required no lining and was reversible. In 1973 shearling (sheared lamb) was used in garments which needed no lining. Sometimes, the leather had a design printed on it. At other times it was worn leather to the outside with the fur acting as the lining.

Many furs are naturally so bulky that they are unattractive except in small-size garments. Evening fox coats, seen in the 1940 movies, were worked vertically with leather inserts to slim and flatten them as much as possible. The new layouts include working these long-haired furs in the round, diagonal, or even old-fashioned let-out. Normally a fox garment is bulky, but carefully chosen patterns, inserts of leather or grosgrain, and combination with other furs and materials help to overcome this disadvantage. Because of this there has been a strong renewal of desire for the long-haired fur garments. Even the fox, long neglected, has returned in coats, trimmings, and hats.

Some furs with low resistance to abrasion, principally those with short stiff hair, are handled with this deficiency in mind. One of the common techniques is the addition of leather, grosgrain, or other material more resistant to abrasion to all edges subject to wear, including pocket edges, cuffs and collar edges. Many garments are also made with the extensive addition of leather or fabrics to lower the cost of the garment. The recent popularity of the battle jacket is an example.

FUR PELT LAYOUT DESIGN

If a single development can be said to typify the fur industry of the 1960's, it was the development of new ways to assemble pelts into garments. These techniques have created a potentially broader base for fur merchandising. There is every indication that this trend has only begun.

Only 20 years ago the fur pelt was usually worked "hair-up" with only occasional ventures into other layouts. As early as 1920, muskrat and ermine manufacturers used diagonal and chevron skin layouts. Natural muskrats were split into five sections, the dark-brown back about four inches wide, two side "golden" sections each about one-and-a-quarter inches wide, and two silver bellies. The backs were zig-zagged together and the gold and silver sections joined in a "v" seam, sometimes using a shingled seam. These never attained any large scale of popularity and were considered novelties. Probably, they were too advanced for the times.

The difference between these earlier novelties and the present time is that today novel and unusual fur garments are the norm rather than the exception. The creative advantage of these new developments is that they do not require a thorough knowledge of fur pattern design. Any imaginative stylist can develop ideas by working with a given pattern and available skins or assembled plates.

Summarizing the new trends that have developed:

1. There has been an increase in the

utilization of readily available although heretofore not widely used furs, as well as a concentration on rarer, more exotic furs.

2. Closely related has been the employment of plentiful but never-before-used furs. Examples include African antelope, kangaroo, water buffalo, zebra, and several varieties of sheep. Many of these have recently become usable in fur garments because of advances in fur dressing which can make them thin-skinned and lightweight.

3. Fur pattern styles have been adapted for the younger set, not only in line but in size. Fur houses now stock sizes 4, 6, 8, and 10 which, except for a few specialty houses, would have been made-to-order only 30 years ago.

4. Original combinations of furs are used. Some high-fashion houses may have carried trimmed seal or mink with fox or sable in the past, but it was not until the late 1960's that these combinations became a standard item in the good, moderate-price lines.

5. Furs are combined with non-fur materials. Leather and fur combinations in which the leather is an integral part of the garment, not just trim, is a trend that is certain to spread to other materials.

6. The use of assembled (pieced) fur patterns has increased. Previously there were a few standard, limited, and long-established monotonous patterns restricted to ovals, gills, sides, etc., in mink and similar bodies. Now, sheets of pelts, paws, ears, heads, sides, etc., have been introduced in other furs. This important phase of the fur industry blossomed into multi-colored designs, tweed and figured effects and even combinations of one fur with another (mink with Alaska seal pieces).

7. By scientific breeding animals in new natural colors are constantly being developed. The fur-dyeing industry is also constantly providing new effects and shades by experimenting on standard pelts. Natural green, red, and black mink have not been bred yet, but such colors are readily available in dyed form. Our avant-garde designers have not only made whole garments from such colors, but have used them as trim or even as inserted designs on other shades, natural or dyed.

TYPES OF FUR GARMENT STYLES

Most of the descriptive names used to delineate fur garments are derived from or can be used with equal validity in other apparel fields. Descriptive millinery terms such as toque, cloche, tam, overseas cap, and homburg apply to non-furs as well as to furs. It is pointless to try to determine in which field the names were first used. The following garment style names are confined to or are more common to fur than other apparel materials.

Choker Scarf: Made from fox, mink, marten, sable, etc. It includes the paws, tail and ears. The head is rebuilt by the insertion of a head form, usually of rubber, and the eyes are reproduced in glass. This type of fur ornament was popular until World War II. It was worn over the shoulders in carefully arranged sets and held in place by ball or spring clasp snaps

The wide variety of small fur pieces has grown immeasurably in this century, in part due to the strong interest in owning even a small fur in mink. Here, two basic cape styles: (l-r) cape jacket, semi-fitted in front; and the clutch cape.

Fur jackets and shortcoats have also been important since 1950. An example of furs for leisure pursuits is the parka at left. At right, the single-breasted jacket with a single row of buttons down the front.

mounted as part of the jaw. This fashion may be said to be the descendant of the original fur adornment. The return of this item is problematic so long as conservationists are active because the very form of the garment is an affront to their program.

Boa: A modern modified version of the scarf made from one or two skins. It is approximately 6″ by 40″ when finished. It has a roughly rectangular form with a slight curve to make it fit around the neck and the shoulders. The underside, which for the most part is not covered with fur, is composed of wide grosgrain covered with velveteen or other fabric. A spring snap is mounted underside at both ends. Either or both ends can be snapped to each other or to the loose edges of the

velveteen, making it possible to wear the boa in several different ways.

Fling: Long, tapered thin scarf, with or without tail-trim edge. Most popular is mink. Most flings are 72″ to 80″ long, although some have been made as long as nine feet. Their width varies from about four inches at the neck to eight inches at the bottom. They usually taper both ways from the neck.

Shrug: Small, short shoulder cape, elbow-length or less, held in place by shaped shoulders and by one or two front fasteners. Popular in mink and in blue and white fox. Also made in white rabbit.

Sling Cape: Longer cape, approximately jacket length. Designed in many variations, from simple straight line with slits for hands to elaborate fronts and side

The right collar treatment has often been key to a fur garment design. In formal wear, the cowl collar (l) has been a standout. Another favored look has been created with the use of the notched collar, at right.

Many consumers prefer simple collar treatment. Among the more conventional styles are those provided (l-r) by the peter pan and shawl collars. Mandarin, choker and tuxedo collars are others developed to enhance fur styles.

The fur stole may be the best known of the so-called smaller fur pieces. Stoles cover the shoulders (left), at times double as muff. Choker (right) keeps the neck warm. Usually, it's a single well-furred skin.

Two important neckpieces are (l-r) the ascot, or tie neckpiece, and the boa. Ascots generally tie diagonally, as shown. Boas are usually softer, longer, narrower and fluffier—a type of scarf in fact.

Since fur lacks the suppleness of fabric, sleeve treatment is a special challenge to fur designers. Two designer achievements are (l-r) the balloon sleeve and dolman sleeve. Rounded effect is key to the balloon; dolman gives capelike effect.

Back treatment has become increasingly more important in fur coat designing. For high-fashion customers, the rounded and full backs (l-r) are among the most popular. Interesting contrast here shows rounded back tapering at skirt, while full back flares.

---

shirrings which give the appearance of a sleeve.

Bubble Cape: Waist length cape, usually 18″ to 21″ long. Many versions have *cabuchon* or a cupped effect at the edge, which is gathered. The front is the same as a short-suit stole. The main difference is that the skins are worked vertically.

Clutch Cape: Same general dimensions as bubble cape. Distinguishing characteristic is in the design of the front. Collar usually extends into the tuxedo effect with pockets at lower edge, enabling wearer to insert hands and hold garment in place, often by crisscrossing fronts. The non-tuxedo version has a slit pocket.

Bolero: General term for several variations of the cape with elongated fronts. Characteristic construction includes horizontal stripes across the back curving to vertical on the front sections. May be an almost straight "contour" stole, with or without a collar. Front extensions vary in length and may be rounded or squared-off at the ends. Front extensions have slit pockets which may be used to hold the stole in position. Almost always made in mink.

Trotteur: Fingertip length walking coat, about 32″ long.

The set of sketches which accompany this chapter are from one of the Fur Information and Fashion Council's many merchandising booklets (see bibliography). It shows many of the major apparel styles and terms common to the fashion industry and the fur industry.

BASIC FUR PATTERN TERMS AND FITTING ESSENTIALS

The total outline or appearance that an item of apparel presents is its silhouette. It may be "straight line," "fitted," "flared," "A-line," etc., in accordance with the visual effect it gives. Complications arise during one of the cyclical fashion transformations from a simple straight-line silhouette to a flared and fitted style. Assuming that both garments are made from the same number of skins, the complications in manufacturing the latter add substantially to the cost of the coat.

A typical fur pattern for a simple, classic, narrow silhouette garment with straight sleeve, cuff and collar looks approximately as shown in the accompanying diagram. A normal fur pattern usually represents half of the garment, unless the style is asymmetrical, or is a custom-fitting for a special figure problem.

Without delving into the technical aspects of how a fur pattern is created, they are generally produced from a basic

ELEMENTS OF A FUR COAT PATTERN

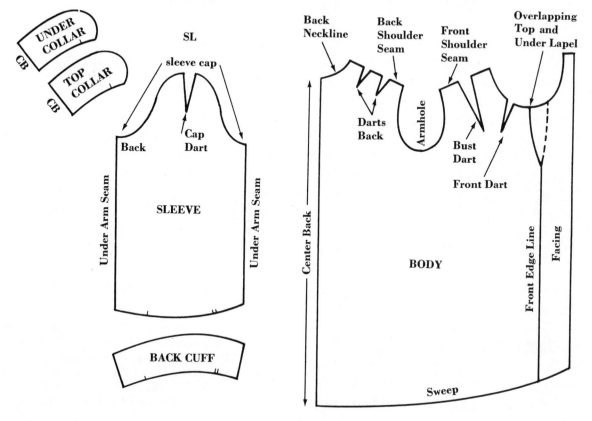

elementary pattern commonly known in the dress trade as a *slooper*. Other times it is created from a paper reproduction of a cloth design or *drape*. In both cases, the following factors must be considered in creating different areas of the fur pattern:

The Body:

1. Length. In most instances furriers include an allowance for turn-up of one inch, unless specifically marked *for full length* as for a French bottom without any fur turn-up. A 41 in. garment would be cut to a 42 in. length. A fur coat is usually made one inch longer than the dress length, since most fur is normally less supple than dress material.

2. Overlap. The overlap is the part of each front which extends over the other. The overlap is normally about two to three inches at the sweep. It may maintain this width all the way up the front, or narrow or widen slightly, depending on the style. The only exception is the garment with a closing at the center front, for example, a zipper front. Most customers want a generous overlap in their garments so that the front doesn't open at or below the knee when they walk. The trouble is that over-generous laps are droopy and clumsy when worn open. They also tend to make the garment look bulky even when closed. The customer will often misinterpret the fitter's honest desire for a trim, good-looking front as an effort to save material.

3. Fit. Bust and Shoulders. Normally, the most difficult of all adjustments to reconcile with the ideas of the wearer is the fit around the armhole, bust, and shoulders. These disagreements invariably involve the customer choosing a close fit, as opposed to the fur fitter's knowledge of the ease required in a fur garment, which cannot be made to stretch and give as can most fabrics. To add to the complications, recent government statistics indicate that women are getting bigger all over. Style sizes have not kept pace. The National Bureau of Standards reports that the measurements of a size 12 in 1971 are 35-26-37, as compared to 34-25-36 in the 1940's, a change unnoticed by manufacturers.[1]

4. Silhouette. Most fur fittings are made with the customer standing still. A narrow silhouette that barely clears the posterior bulge becomes all too revealing when the wearer moves.

Armhole and Sleeve Cap:

A common misconception held by women is that in furs, the larger the armhole, the more comfortable the arm movement. A semi-dolman sleeve (one that ends half-way between the armpit and the waist) is thought to give more ease than a tailored, smaller armhole. If this were true, the full dolman (ending at the waist) would then be even more desirable. For slight arm movements, loose sleeves are comfortable. However, when lifting the arm, this heavy sleeve pulls up the entire garment. Therefore, the very large

Is this chinchilla coat perfect? ▶
Constantino Christie of Christie Bros. confers with Designer Henri Bochene (with tape) in typical design situation. Concern here is real. Much is at stake!

64

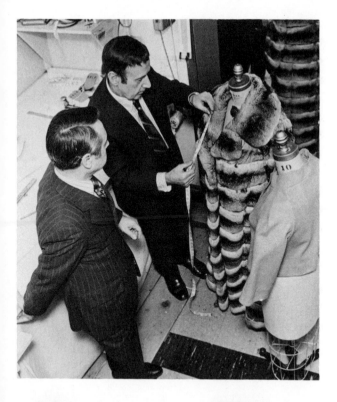

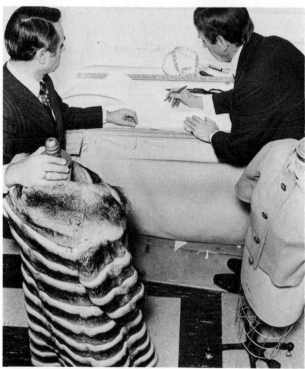

armhole is undesirable. By the same token, the very small armhole is also undesirable because it is uncomfortable and has a lack of ease.

The Sleeve:

1. **Style.** Over the last half-century, styles have varied from the voluminous bell, dolman, and cape sleeves to the narrow, tailored, or straight sleeves. The latter have maintained present popularity because of light weight. The smaller the armhole circumference, the more likely that the sleeve armhole has been constructed slightly larger than the body armhole into which it is set and must be eased into. In a very small and tightly tailored sleeve, the elbow measurement becomes a factor in locating the elbow dart or kyle in its proper place.

2. **Length.** Sleeve lengths vary with style trends, but consideration should be made for such occurrences as the wearing of bracelets, which will rub against and wear away fur.

Cuffs:

Sometime ago no fur coat or jacket was considered complete without a *self* or

contrasting fur cuff. These were made so they could be worn long or turned back. The modern desire for slimness and elimination of weight and bulk has done away with the fur cuff on many styles, especially for younger wearers. In the long run, cuffs come and go with the whims of fashion.

The Collar:

1. **Types.** Without going into detail about each type of collar, a few pertinent statements can be made. The fur collar used as a trimming on non-furs must be separated from the rest. The former collars are attached to cloth garments, usually by hand-sewing, making this segment of the fur industry a completely separate area. The makers of these collars are called trimming manufacturers, and usually specialize in the manufacturing of collars and cuffs exclusively. They have their own association, and are listed separately in the fur industry directories.

2. **Styles.** Each of the common fur collar styles has its own advantages. The personal taste of the wearer should decide which type of collar style is most suitable. Some styles are more utilitarian than decorative and can be worn only one way. Others can be worn in two or three ways. The wisest decision should be made by the wearer in accordance with her taste rather than the current style trends. The selection of a collar style dictates the size and contour of the neck opening. The neckline compatible with a shawl collar differs radically from one that will accept a notched or portrait collar. The wearer who desires warmth should consider this carefully. For women who can't stand confinement around the neck and head, there is a group of styles characterized by a set-back or saddle. This is a narrow band of fur, one to one and a half inches wide, which fits around the top of the shoulder next to the neck, starting from the center back and coming to a point in front. The collar is attached to the outer edge of this saddle, setting it away from the neck and allowing the head more freedom of movement.

CONCLUSIONS

Time and time again, fur fashions have shifted from one extreme to another. In 1970, the midi (mid-calf) and the maxi (ankle) lengths vied for the season's supremacy against the prevailing mini length of the past few years. Logic would dictate that these longer styles, although not heavy and bulky in cloth, would be impractical or unsuccessful in fur. In the early 1970's, however, these styles have become very popular. However, there is no guarantee that full-length garments will not return to the 1949 Dior lengths, or the short thigh-high hem.

FOOTNOTES—CHAPTER 5

1 "The Girls are Getting Bigger All Over," Ann Blackman, *Ft. Lauderdale News*, Jan. 8, 1971, p. 18.
2 *Assistant Fur Fitter*, Sol Vogel (New York, Sol Vogel, Inc.).

6

The Manufacturing Process

6. The Manufacturing Process

1. Assorting:

The garment manufacturer usually buys his raw or dressed pelts by auction. If the situation warrants, he deals with an intermediary, the skin dealer, who buys very large lots at auctions or buys privately from other sources and breaks the lots into more marketable units. The skin dealer may also have the pelts dressed if they are raw. The skin dealer will finance purchases over a period of time, if necessary. The furrier, if he cannot afford to buy at auction, buys exactly the quantity he needs from the dealer. If the furrier is a retailer, the purchase may be only a single bundle for a single garment.

The traditional functions of the skin dealer, including large-scale buying, dressing, assorting, and financing, show signs of losing their importance. Auction company skin specialists often do the assorting themselves, which has tended to reduce the skin dealer's functions to a broker's, to clearing a purchase and extending credit.[1]

Assorting, however, is an important service the dealer still performs, especially for the retailer. A given lot purchased by a dealer is often from a single area (if wild) or from the same ranch or herd. One would therefore assume that the live ani-

The fur grader is creating bundles of mink skins based on color, quality, fur density. Uniformity is the keyword.

69

mals were probably blood relatives and that the skins would closely resemble each other. If this were true, any sampling of the skins could be put together into a well-matched bundle for a fine garment. In reality, however, this is not so. Animals, although coming from the same relatives, may be as different as children from the same parents often are. They will share many features, but will still appear as distinct individuals.

The proper arrangement and grading of the skins into *bundles* (the number of skins necessary to make a given garment) is a skilled process and one of the most important steps in the production of a fine garment. This bundling service is supplied, often in advance, by the dealer or one of his employees who is a qualified assorter.

The contrast is striking between the attitude of the industrial American rancher who prefers to sell his product under his own label or under the umbrella label of his association, and the less personalized attitude of the Scandinavian rancher. Scandinavians have no objection to *intersorting*, that is, to mixing their own skins with similar grades of skins from other ranchers. The auction-house assorter, having a greater number of skins with which to work, is able to assemble larger, more uniform *strings*, or matching fur bundles. The individual rancher is paid by the numerical proportion of his skins in the string when they are sold. The

These bundles are on their way to becoming mink garments. Each bundle will be converted into a coat or jacket.

buyers, of course, prefer these more uniform lots and may pay more than they would otherwise.

In the U.S., after the bundle has been purchased, the final assorting of the lot into garment bundles is usually done in the factory. This assorter may be a principal of the firm, a key employee, or a specialist who is retained to do this work. He will be paid for this work by the piece, so much per skin. His experience and skill is sometimes limited to a few varieties of skins.

The number of judgments the assorter must make on each pelt are astonishing. Among the factors he must consider are: size of pelt, height of hair, density, color, clarity, glossiness, and absence of faults. He must also consider the number of skins in the lot, how closely each skin matches the other skins in the same lot, the types of garments to be made, the average price range that these garments will be sold for, and the types of customers that will purchase the garments.

Because of the disinclination of today's youngsters to go through the long apprenticeship necessary to learn this highly paid craft, attempts are now being made to mechanize and automate some phases of assorting. Much of the needed technology exists in other fabric fields. Whether this technology can be modified to handle furs with the great differences that exist among pelts is as yet unanswer-

Moment of truth begins: For cutter is now laying out skins from bundle to make certain they all match.

able. Given the vast number and kinds of judgments that the assorter must make, the problem becomes most formidable.

2. Stretching and Preparing:

In assorting, the pelt has not been physically altered beyond the preliminary dressing and processing necessary to make it ready for use. Next, a work ticket is assigned to an appropriate bundle of skins and the actual process of making

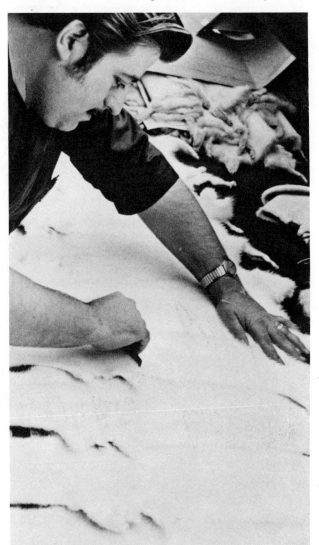

the garment begins. For most types of pelts, the bundle is turned over to a *stretcher*, a floor worker or a specialist who devotes all or part of his time preparing the skins for the cutter. If the skin has been *case-handled* (dressed without opening the belly) the stretcher will open the pelts, make some preliminary opening cuts, and then wet and stretch them. Sometimes, the cutter will stretch the pelt himself.

For most furs, however, the stretcher soaks the skins and, with his hands or a metal bar stretcher, works the pelt into the maximum size and shape that the cutter wants. He may also do some preliminary trimming of the skin, cutting away the tail, opening the paw pocket, or trimming away the head, depending on the variety of the fur and the use to which it will be put. Usually, the stretched pelt is allowed to dry overnight before it is turned over to the cutter.

3. Fur Cutting:

In the hierarchy of the fur garment manufacturing business, the cutter is supreme. He is held responsible for a garment's deficiencies in fit, appearance, and so forth. It is his responsibility to make sure that each skin in the garment relates artistically to its neighbors, creating an attractive composite. Each pelt must occupy its assigned space and shape in the whole, no more and no less.

Because the assorter has assembled the bundle with great skill, it might be assumed that the artistic part of the cut-

The cutter has to eliminate waste. Skin at bottom shows the furskin before trimming. Above, the cutter has done his job.

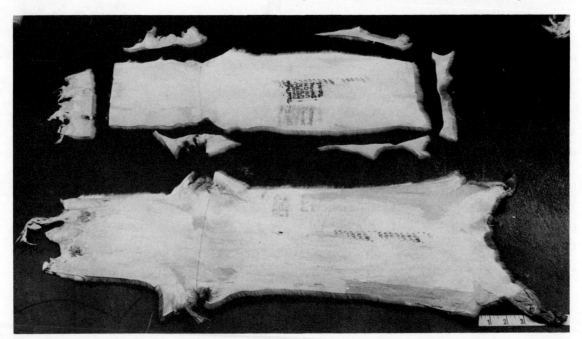

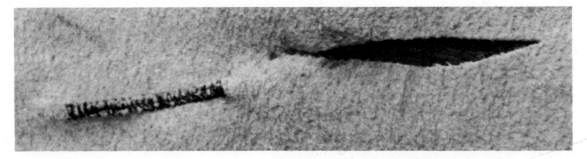

Getting rid of damages. This blow-up indicates two techniques: at left, leather is sewn; right, damage is cut out.

ter's responsibility is minimal. Unfortunately, the nature of most furs, dyed or natural, sheared or uncut, is not that uniform. Even within the carefully assorted bundle, some variations of size, color, height, density, and glossiness, remain. Given a bundle of skins, a good cutter will match out a garment whose eye appeal makes it far more valuable than can a less experienced cutter handling the same bundle. Therefore, a less skilled cutter works on the lower-priced furs where the level of artistic skill required is not so great.

The problem of reconciling shade, gloss, height of hair, size, and other factors in a comparatively simple garment, such as an eight- or ten-skin stole, into the most harmonious sequence possible may take more than half the time the cutter spends on the garment.

The second, less important phase of the fur cutter's work is the actual cutting of the skin. He uses a fur knife, a device consisting of a special holder into which he places half of a double-edged razor blade. The difficult technical problem is to cut the leather without cutting the hair or fiber. This is accomplished by barely penetrating the leather with the tip of the knife blade and placing minimum pressure on the hair and fiber base.

Another part of the cutter's job is to trim the fur skin for use in its assigned place on the pattern. The trimming of a skin is not merely a standard mechanical operation always done in the same way with the same species of pelt. On the contrary, it is sharply modified depending on the use to which the pelt is to be put. If the skin is to be used for a scarf, boa, collar, stole or coat, a variation of trim is required in each case.

As he trims the skin, the cutter examines it for faulty areas or blemishes called *damages*. Depending on the species and grade of skin, there will be variations in the number and type of such faults. In every skin, the presence or absence of damages is a major factor in determining its grade. In some species of pelts any substantial damage makes it commercially useless. In others, damages are less disastrous. Fortunately, almost all damages can be repaired. This adaptability to adjustment makes it possible to repair and remodel anything in fur, from a skin to a whole garment.

The fur cutter has several techniques to fix damages so they will not show. One might assume the most common method is *patching*, inserting another piece into the damaged area. This, however, is not true. Patching is usable on lamb but on almost no other fur because animals are so different from each other that it is almost impossible to match perfectly a piece of one animal into a section of another. In lambs, especially the persian lamb with its curled fur and dyed black or brown color, the method of patching is very successful and widely used.

For a small damage, about ½" wide, the common technique is to cut away the substandard area in a long pointed oval called an *eyelet*. A seam is then made to close it, and the natural elasticity of the leather eliminates the lost area.

A larger damage, too big to be handled by an eyelet, presents another problem. Since most furs cannot be patched, because the added piece will show, the missing area can be filled in only by the fur from the adjoining area of the same skin. If any single technique can be said to make fur garment manufacturing possible, it is *dropping a tongue*. Within limits, any damage can be handled by one of the forty or more variants of this technique.[2]

Use of tongue technique to alter shape of a skin. Note how fur is cut to create more width as well as new shape.

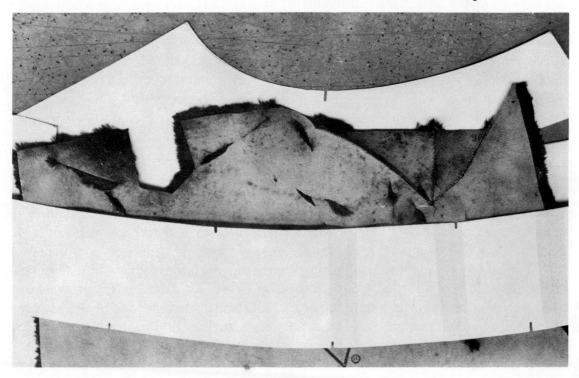

Properly cut and sewn, the *tongue* will not show on the hair side. It must be emphasized that the presence or absence of damages is not an infallible criterion of the quality of a pelt. In some species where such faults are easily corrected a skin with one or two damages may have a superior hair or pattern quality to an undamaged skin and so be considered more valuable. On bold-patterned skins, however, a damage in a prominent place may reduce its usefulness and value. (Many variations of the tongue technique are used to alter the shape of the skin to fit pattern requirements and have nothing to do with damaging.)

The trimmings and cuttings are not considered waste, especially those from the persian lamb, fox, seal sable or mink, whether they be the heads, tails, paws, gills, sides, or ears. In the United States they are carefully collected and sold to exporters who assemble them in large quantities.

How will the cutter combine, assemble or join the fur pelts? As we have already suggested, the development of varieties of layouts is one of the major advances in modern fur garment manufacturing. Following is a summary of some of the principal systems now in use.

Block Layouts:

This is one of the oldest and simplest method of skin joining. Originally, the skins were cut into simple rectangles,

Block layout: this is one of the oldest methods of joining skins. Probably used by ancient man in sewing his clothing.

with the hair flow up, down, or horizontal. Elaborations later developed, involving diagonal layouts (ermine, fitch), octagonal (muskrat), diamond (muskrat), and split skin (kidskin, fitch and fox). Wherever the shade of the pelt permitted, rounded head-to-rump joinings (natural raccoon) or the "V" (squirrel) saved material and increased pattern layout interest. Whatever the variation, and there are many, the distinguishing characteristic is that each skin unit can be discerned and counted.

Concealed Seams:

The cutter's goal for certain furs is to produce a line of skins in which the individual boundaries of any single skin cannot be detected on the hair side. When well made, the line or row of skins appears

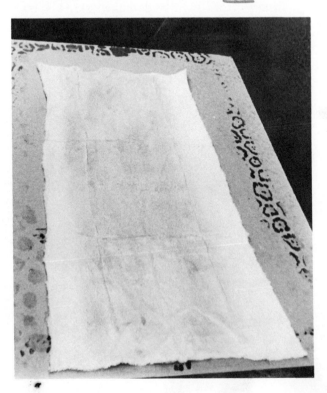

to be one sheet of fur. The best example of this technique is black persian lamb. This technique is also common on lambs, sheared rabbit, and some seal. When black persian lamb is examined on the leather side, the skins are found to be joined by a form of irregular or uneven seam which may take the form of zig-zags (large saw-tooth), scallops (waves), or a completely patternless joining. The furrier has used the principal of camouflage to hide the joinings of the pelts. The broken and irregular type of joining tends to confuse the eye, which is trained to recognize only regular shapes. As a result, the eye "does not see" the joining.[3]

Garments made from pelts in the lamb family seem to have many patches when viewed from the leather side. This might seem to indicate that the garment was made from greatly damaged skins. However, these pieces are sections which the furrier has added to improve the appearance of the fur and are called *flowers*. Flowering is considered a sign of quality production. Because this type of concealed patching is most adaptable for use on lamb, these furs are the most readily repaired and remodeled.[4]

Assembled (pieced) Furs:

Five hundred or more years ago, venturesome Greek sailors brought back from the East samples of fur and fur garments. These were assembled from left-over waste cuttings of skin garments. The Greeks adapted and refined the technique, importing fur cuttings from the world fur centers and assembling them into *plates* or *bodies*, a sheet of fur material made up of many small pieces of fur.

The world center of the assembled fur piece industry is still the Greek city of Kastoria, although this type of fur artistry has spread to other countries. Skin garment manufacturers carefully save all the

Joining of skins: left is zig-zag technique; right the scallop method. In both cases, aim is to conceal the seam.

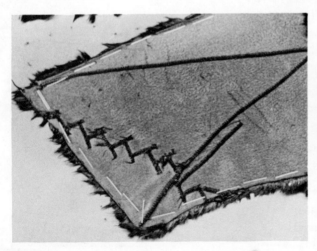
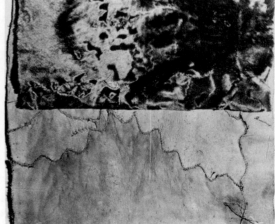

The mink plate—important manufacturing impact. The variety of design opens up new fashion vistas in fur.

cuttings in their fur garment manufacturing process. These cuttings, with the possible exception of the tails which may be sold separately to fur hat makers, are sold by the pound to piece dealers. The piece dealers sort out these pieces, then bale and ship them to purchasers in Kastoria or other plate-making centers, usually in large bales weighing more than 400 kilos or nearly half a ton. The most popular pieced garment is mink, with persian lamb paw second. In working out mink cuttings, the pieces are first sorted by type: heads, front paws, rear paws, gills, sides, and so forth. If the shipment contains intermingled colors, they will be sorted, although this is usually done before shipping.[5] The sorted pieces go to the cutter, who cuts them into the required shape, while slicing away unusable parts. The ingenuity which has evolved in shaping the piece so that it can be sewn into an attractive pattern is evident in any good assembled fur piece plate. The cut pieces are passed to the operator, who joins them in an endless *string* or some form of patterned panel.

The strings or patterned panels are worked out into 40-inch or 48-inch lengths, depending on their destination.

The sewn lengths are again matched so the plate will be uniform, or at least shaded off artistically. The matched strips are then joined into *bodies*, a term referring to a unit from which a full length garment can be made. Finally, the sewn plates are nailed (blocked) out so that they lie flat, with uniform lines.

The plate is handled just like cloth yardage. The pattern is carefully placed to maintain layout symmetry and balance. The rest of the production follows much the same steps as on any other fur garment.

Fur Garment Sewing:

The first fur garments were, of course, sewn by hand. Hand sewn assembled skins or piece plates are still marketed today, coming from undeveloped areas of the world in parts of Asia and Africa. Eskimo tribes also tan and hand sew their own fur garments.

The advanced nations use the fur sewing machine, a single-needle, single-thread unit (foot powered where there is no electricity) in which the needle moves horizontally through the fur creating an overlock stitch which is closer to a knit than a seam. The fur operator's work is by far the most exacting in the field of apparel machine sewing. He works with an elastic material which varies in thickness. The tolerance of error is less than 1/64 in. As he joins any two pieces of fur, the operator must tuck the hair into the hair side so that none will be caught in the seam and show up as a damage on the fur side. He must also exercise considerable judgment as he joins and shapes

each seam to execute what the cutter marked out. In addition to working out the shape of the individual skin, the operator joins the skins together to form the major parts of the garment such as sleeves, body, and collar. Sometimes, the skins are joined together directly. At other times, they are divided with strips of leather or grosgrain. In the United States and Greece this work is mainly done by men. European and Asiatic fur operators are mostly women. It now takes from six months to two years for a beginner to become a productive operator.

Let-Out:

The leather side of a *let-out* fur garment appears to be made of hundreds of narrow, diagonal strips sewn together. In reality, each section originally was a full skin. The cutting and sewing together of each of these strips less than ¼ in. wide, constitutes one of the most difficult forms of fur craftsmanship.

This technique enables the furrier to lengthen most long-haired furs and some others (such as gray persian lamb) with a minimum loss of area, and also to preserve the coloration proportions on the hair side without the work showing. The ability to lengthen the skin enables him to extend the skin in one unbroken line from neck to sweep, eliminating cross seams or joinings. Artistically, this uninterrupted flow has a slimming, flattering effect on the wearer. The narrow lines accent height and minimize bulkiness.

The elongation of the individual pelt also accents and enhances the change in color from side to side of each skin.

First step in letting out process. Cutter slices mink skin into narrow strips. Top cutters are highly prized.

Curved effects are possible. The elaborately curved fronts and shaped collars and cuffs seen on high-fashion fur and cloth garments would ordinarily not be possible without the let-out technique. Narrow band effects would have to be made from joined sections of skins, instead of the full skin elongated and shaped to the dimensions wanted.

For these and other technical and artistic advantages and because it is so obviously a status symbol of costly workmanship, the term *let-out* has received general acceptance by the public as a sign of quality in a fur garment. Many uninformed customers ask for it on furs which do not lend themselves to this technique.

The originator of the technique is unknown. It was in general use after World War I. It is a logical extension of the tongue method of skin repair and adjustment previously described. The

Stages in letting out process. Top, trimmed skin. Center, skin sliced. Bottom, skin has been sewn together.

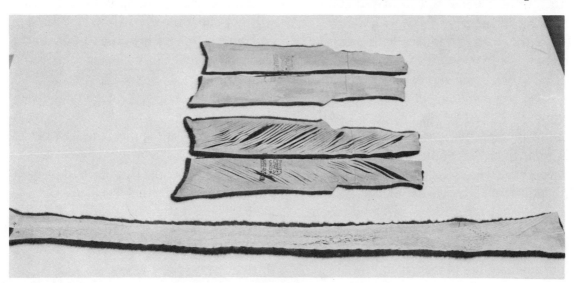

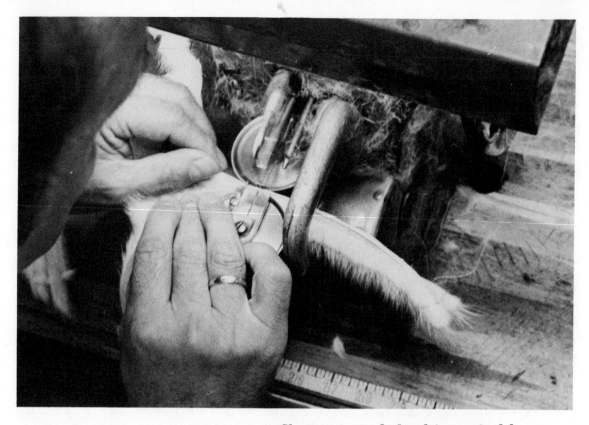

Sharp eye, steady hand is required by operator who is resewing mink strips. Results shown at bottom page 79.

basic principles are shown in the accompanying diagram.

Some basic facts about the let-out technique follow.

1. There is no appreciable loss of area because of seaming, although anywhere from 10% to 20% area loss occurs in the entire process. The machine needle used is finer than the thinnest silk pin made. The thread used is almost as thin as human hair. The tension of the seam is so tight and the stitches so close (about 30 to the inch) that even when rubbed flat and stretched, the edges of the leather of each seam hold together tightly.

2. If correctly cut and sewn, the seam will not show on the hair side. The skin will appear to have grown naturally in the shape it occupies in its assigned place on the pattern.

3. Some small percentage of the elasticity is lost.

4. There is no appreciable loss of durability.

5. The type of thread used usually outlasts the fur leather itself.

6. The greater the number of cuts, the higher the quality of the work, that is, the less the seaming is likely to show on the hair side.

7. Most let-out furs show their cuts under special conditions, such as over a curve of the body, at the sides of the skin, when held upside down, or when left uncleaned too long.

Nailing:

When the operator has finished, the assembled garment is in its major parts—body, sleeves, collar, and so forth. Each part is, however, wrinkled, puckered, contains unwanted folds, and in general does not lie flat, nor exactly fit its pattern. Eliminating these irregularities and making each part fit its assigned place on the pattern is the nailer's job. He marks out the pattern on a paper-covered nailing board. Formerly this nailing board was made of redwood or other water-resistant wood. Today, the furrier-nailer has substituted a board made of pressed paper. The leather of the sewn and assembled parts are thoroughly brush-wetted with water. These sections are then stapled to fit the marked out pattern. When the wooden nailing boards were used, the nailer would tack out each section utilizing special nails and a fur-nailing pincers. The nailing board made of wood is still used, especially for nailing the thick-leathered furs, but even here it is rapidly disappearing. The use of staples is faster and gives better results. In the United States the open staples are worked by

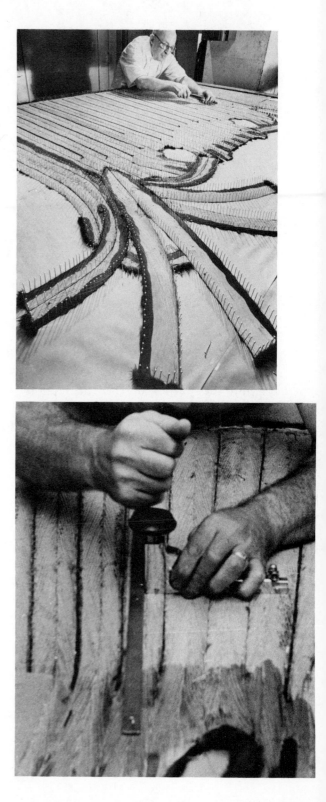

Literally thousands of nails are used by the nailer on the leather side of this coat to establish the exact shape of the final garment. At bottom, he uses staples to block out lines on coat. Metal strip helps keep lines straight.

hand, but in Europe they are often applied electrically.

The work of the nailer would be comparatively simple if all he had to do was to pull the leather of the fur section into shape so it would fit the pattern, thus eliminating all folds and wrinkles. On sheetlike mono-color furs, such as black persian lamb, this is almost all that is needed. However, whenever there is any sort of vertical, horizontal, or diagonal pattern in the fur assembly, the nailing problem becomes far more complicated. The nailing must follow the positions as planned on the pattern, proportioned and symmetrical. Any failures to set them into their assigned positions will show up as obvious faults in the finished garments.

The nailed garments are allowed to dry overnight or longer if necessary. Other supplementary processes are sometimes applied on the dried, nailed garment sections. If the leather color is markedly lighter than the fur color, it may be necessary to darken the leather

Stapled out seal coat, above. Below, use of nails to block out the seal coat. From leather side, it's hard to see fur beauty.

82

color in certain areas by dyeing the leather to match the color of the base of the fur hair or fiber. The purpose is to hide this color contrast so that the leather will not show through when the fur fiber is parted. No change is made in the color of the fur hair or fiber. This technique is known as leather *tipping* or *staining*, and is not included in the process which alters the color of the fur itself. A tipped garment is still a *natural* garment if no other coloring process is used on the fur.

Squaring:

After drying, the garment is for the first time ready to be cut exactly to the pattern. The pattern is carefully and accurately marked on the dry fur leather which has been stapled slightly larger than the pattern. No seam allowance is needed or allowed for. Each marked part is then trimmed exactly to the marked line.

Taping:

To prevent unwanted stretch, the edges of the squared parts are, in most furs, held in place with tape. Friction tape is applied to the edges of the fur with the tip of a controlled heat iron. On luxury garments, many furriers hand sew a flat tape to these edges in the finishing department. The taped garment parts may, at this point, receive additional treatment designed to re-fluff the hair, perhaps still matted from nailing. On Alaska seal, for instance, a high-power steam gun is used. The necessary joining points may also be marked out.

Closing:

Like many operations in fur, the job of assembling the "squared" taped sections of the garment into a unit is designated by a name which bears little or no relationship to the function performed. The operator who does this work at a fur machine is known as a *closer*. The fur edges he joins are stiff and thickened by added tape, so he uses a needle and thread one or two sizes heavier than that used to sew the garment. The seam sewing itself is the least difficult part of his job. He is not required to make a very thin seam. His main problem is to see that no fur is caught in the seam and that it grips both sides firmly. He is not expected to work at the speed of the operator.

The important consideration in closing is correct joining and accurate fit. With each part carefully marked for joining, this might seem comparatively easy. However, in practice, the complexity of anything beyond the simplest pattern has proved to be beyond the capabilities of most operators who would like to move up to the higher-paying job of closing.

Accurately joining two supposedly equal taped edges can become a serious problem on a fur sewing machine whose two material flow controls do not operate as a pair or on the same drive. The inside wheel is powered like the rear wheels of most automobiles, while the outside wheel is not powered, but operates by friction, with the fur leather between. If not carefully controlled, the inside edge of any fur seam will feed through the machine faster than the outside, throwing joining marks out of alignment and creating unwanted

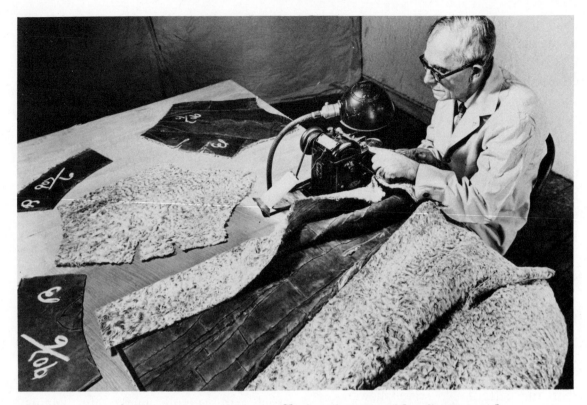

Closing time: as with a jigsaw puzzle, this sewer is joining body, sleeves, cuffs, collar of a persian lamb coat.

curves. On the other hand, when the pattern calls for it, pleats, folds, darts, and shirring must be placed with the utmost accuracy and without the assistance of the special attachments available on other sewing machines. Unless the closer has a clear understanding of how the parts of the pattern are to fit together, much time and effort will be wasted.

The ability to visualize the assembled parts as a complete unit is not common and is the main reason good closers are in demand. Certain other minor preparatory steps complete the closer's work. Before being turned over to the finishers for lin-

ing, the garment may receive additional attention such as inspection, pressing, touching-up, and drumming to improve its appearance and fit.

Finishing:

Finishing refers to a separate department of the factory in which lining, underlining, fasteners, pockets, pads, edging, and almost all the cloth parts are added to the fur garment. If there are five or more workers in the department, a head or foreman finisher usually manages the division.

If it is a custom garment requiring a fitting, the closed garment will be prepared for the try-on by *working it around* or preparing it. This means that a minimum basting of edges, pads and pockets, cuffs and fasteners will be hand sewn into place. For most garments this will include a temporary *silesia* lining to keep the garment smooth, the customer's clothing clean, and the fur leather seams from catching on clothing and jewelry.

The final finishing procedure for most garments can be broken down into two stages. First, the lining and underlining is cut and sewn by machine into a shell. The lining cutting is usually done by the head finisher. He will do the necessary machine and hand-sewing to complete the shell which, for a sleeved garment, will be a fabric garment in itself, ready to slip into the fur garment and be sewn into place. Before that final step, all the other non-fur accessories such as fasteners, pockets, pads, and pellon stiffening, will have been completed.

The joining of lining shell to fur is done by *feller hands*, so called because they use a hidden or *felling* stitch. The lining and underlining at any edge is always cut larger than the pattern. The excess is bent in or doubled just enough to be comfortable and pinned into position. With a *glover's* (3-sided point) needle, the feller hand sews the underside of the edge fold to the fur leather, catching fur, lining and tape in a way that keeps the seam hidden from view on both sides. When this and some other minor jobs are completed, the garment is ready.

Glazing:

All this handling has obviously not improved the looks of the fur side of the garment. After finishing, therefore, fur garments undergo a final touching up. This formerly was limited to a wetting down of the hair, followed by reversal of the hair flow with a dull-edged stick, and was called *glazing*. Today, the steam gun, steam iron, or a revolving heated iron

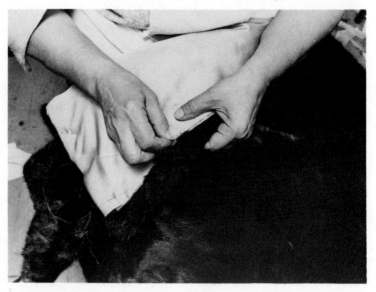

Finishing touch: Fur finisher has to cover all the blemishes on leather side by sewing in attractive lining.

unit, depending on the nature of the fur, is used to restore full fluffiness and uniform hair flow. While this is being done, the garment is also carefully scrutinized for thread ends and damages.

The garment is now ready for its final inspection. It is hung on an appropriately sized figure or, preferably, worn by a live model. The over-all hang of the garment is of primary concern, but other facets of its appearance are also checked. Any damages, as well as pelt alignment, match, fastener placement, and every other visible part of the garment are given close scrutiny. The garment is then ticketed and is ready for stocking or shipment.

Improving and Simplifying Fur Garment Manufacturing:

The necessary technical improvements needed to keep the fur industry in tune with the times have been touched upon in various places in this book. There are unmistakeable signs that such improvements are being developed and utilized in many fur manufacturing centers. If resistance to such improvements on the grounds that they eliminate jobs persists in one geographical area and they are encouraged in another, the outcome may be a shift in importance and sales volume to the more technically advanced area. Time after time, in and out of the apparel industry, a combination of lower wage scales, import restrictions, government subsidies, and the full utilization of technological advances have caused major shifts in world production aggregates. The following is a brief summary of existing and probable improvements in fur apparel manufacturing.

Cutting:

European furriers often use a mechanical fur-cutting machine which cuts all the let-outs into the skin in one motion. It has the advantage of reducing the need for skilled let-out cutters who are among the most highly paid in the industry. The cutters can devote their time to the necessary matching and set-up tasks which no machine can perform. The uniformity of the cuts eases the operator's job and makes for a better looking garment.

Operating:

If Rip Van Winkle had been a good fur operator, and had awakened today after a long sleep, he would have little difficulty in recognizing contemporary fur sewing machines and techniques. There have been some minor improvements, but by comparison with the efficient, high-speed modern sewing machines in other branches of the apparel trades, the fur sewing machine is archaic. Perhaps the small size and great number of firms work against the necessary expenditure of money and time to find new solutions.

There are, however, many possible improvements which would be inexpensive but, for some unknown reason, are never undertaken. The control of an even flow of material through the machine (as needed, for example, in closing) should soon become available. One means to achieve this is a machine with an optionally lockable feed. An improved form of a

Why let out skins? When strips are sewn together, as seen here, they are supple and graceful.

blower vacuum attachment was invented to add to existing models of the fur machine. They are still harder to find than hen's teeth, yet this attachment mechanically tucks in the hair of the softer furs, eliminating the most difficult phase of the operating technique. Preliminary tests indicate that for the furs for which it can be used, it cuts the operator's learning period by 75% to 90%. Improvements such as this may make it possible to replace the fast-disappearing fur operator. The New York industry, with the help of state funds, is attempting to bring this attachment into wider use because of the dual advantages of improved hygiene and easier apprentice training.[6]

There are many other possible improvements which would cost little to institute. The height of the fur machine stand should be made adjustable so that variations in the height of the operators is no longer an obstacle to comfort. The

operator is forced to sit at an angle to the machine with his feet almost at right angles to the pedals. The head of the machine and the pedals should be set at this angle so that the operator can sit comfortably in front.

After oiling the machine much of the oil runs down the sides and often finds its way onto the fur that is being sewn. Directing the oil to flow into the machine would be a welcome improvement. The stitch size and thread tension controls are primitive and have no graduated readings, but must be adjusted by sight or feel. An automatic threader such as is found on most other types of sewing machines could easily be provided.

The fur sewing machine is archaic by comparison with most other sewing machines. This affects the quality of the work produced as well as the cost of manufacture. It is suprising that the industry has been content to put up with this situation for so long.

Nailing:

Equipment now being demonstrated uses a power-driven flow of staples and eliminates muscle power. By varying the size and length of the staples, these machines may be used on heavier furs, which are still being nailed. Staple removal, however, is a hit-or-miss manual combing operation. There has been some use of a tape which is fed out under the staples as they are applied which can be lifted up, with the staples attached, after the fur has dried.

The drying process has traditionally been natural, except when emergencies have required the use of gas-heated drying rooms, fans, or heaters. An infra-red combination nailing-drying unit originating in West Germany was marketed here in the late 1960's and promises to make this process rapid and controlled, rather than leaving it at the mercy of humidity and temperature.

Squaring:

The process of marking and cutting a nailed garment is about the same as it was a half-century ago. The use of perforated plastic patterns through which a suitable powder or marking material can be sprayed or applied, a technique in use in the apparel field, might be used on furs. Also, the present manual application of friction tape should be replaced by a unit to automatically feed the heated tape into position. At the 1969 Frankfurt Fair, sewing machines which fastened tape to the edge of the fur leather without showing on the hair side of flat-haired furs were demonstrated.

Advances in machine shearing of skins and squared sections of garments will probably take the form of those used now in cutting patterns into carpet. A beginning along these lines is the making of grooved patterns on white and other sheared rabbits.

Fur Drumming and Cleaning:

Large wood-lined revolving drums utilizing sawdust or ground nut shells,

commonly found in fur skin dressing plants and garment factories, are traditionally used for cleaning furs. The drawback to using revolving drum cleaning for older, weaker garments is that the tumbling may tear the leather. This danger can be minimized by using individual garment bagging. The ideal solution would be a cleaning system which would eliminate drum tumbling, perhaps limiting the whole process to spray application and vacuuming.

Recent improvements have been made in the fields of cleaning solvent and carrier, similar to home-style pressure can cleaners which deposit a wet powder on the area to be cleaned. The powder dries, supposedly absorbs the stain, and is then brushed off.

Finishing:

The major drawback in present-day fur finishing is the high cost because of the amount of hand sewing involved. Certain fur garment centers have utilized machine-sewing techniques for three sides of the garment, limiting the costly hand sewing to the shoulders and neck, which are the most noticeable. There is nothing sacred about the invisible *felling* stitch deemed so necessary by most furriers. There is no harm in using a visible machine-made stitch to attach the lining to the fur, so long as the thread is appropriate and the seam neat and even. The use of fine blind-stitching machines to attach a silesia-type cloth to the leather side of some light, thin, elastic-leathered furs, to improve the hang and prevent unwanted stretch should be encouraged. This step will not, however, automatically add durability to the garment as a whole.

Conclusions:

Progress through advanced techniques and machinery is usually recognized as necessary and desirable so long as it affects someone else. Once these advances impinge upon the livelihood of a group of workers or an industry, a different attitude is expressed. This fear of the unknown is understandable. Will this result in the loss of my job? Will I be able to adapt myself to the new techniques and machinery? If I accept these advances will other and more "dangerous" ideas soon be forthcoming? These are genuine and painful questions that must be answered.

Some industries, notably the automobile workers and the printers, have accepted and even welcomed technological advances. Instead of fighting advances, they have trained their workers to effectively use the new machinery. The results are obvious. Technology has been used to improve the product, lower the unit cost, and make the industry more competitive. This has often resulted in more jobs and a far greater demand for the product. On the other hand, those industries which have fought the advances have at best waged a holding action. Technological advances will not be forever denied. In this country, industries that have not kept up with the times have found a drop in the demand for their product. Foreign imports have become a problem because of their cheaper price and often better workmanship. Restrictive import tariffs have not had any lasting effect. In the end, the manufacturing centers have moved to the more advanced areas or countries.

FOOTNOTES—CHAPTER 6

1 Sandy Parker, "Fur Trend Could Rock Retail Trade," *Women's Wear Daily* (New York, Nov. 10, 1970), p. 1.
2 David G. Kaplan, *op. cit.*, p. 37–41.
3 Kaplan, *op. cit.*, p. 78–95.
4 Kaplan, *op. cit.*, p. 96.
5 Anna Traianos, "Kastoria's Craft," *International Fur Review* (London, July, 1968), p. 36.
6 "Training Plan May Pay For Anti-Pollution Drive," *Women's Wear Daily* (New York, Nov. 25, 1970), p. 24.

7

Maintaining the Product:
Servicing, Repairing,
Remodelling, Storing

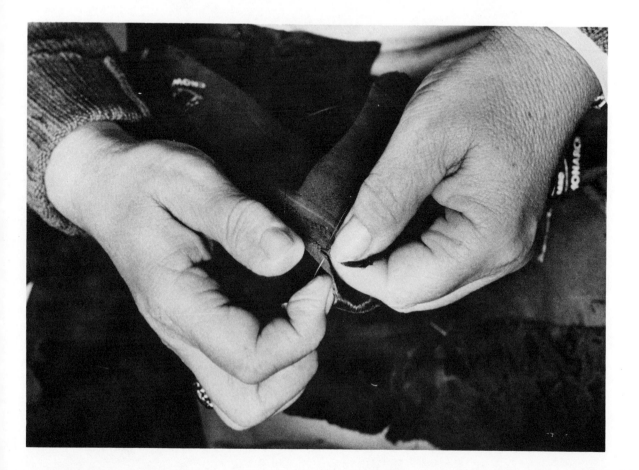

7. Maintaining the Product: Servicing, Repairing, Remodelling, Storing

When considering the relatively high cost of a fur garment, the savings produced by the repairability factor should not be overlooked. A non-fur garment that is torn or has worn edges must be discarded. The fur garment, on the other hand, lends itself to repair. A worn fur garment can often be restored to full use at a small cost. The following is a partial list of the variety of minor adjustments and improvements possible on most furs.

Improving Appearance:

Cleaning: This requires equipment not available to the consumer or the average dry cleaner. Furs are properly cleaned by abrasion rather than immersion. They are not cleaned in the same manner as are fabrics and should, therefore, be sent to the professional furrier for this service. If a fur is cleaned in the same manner as a fabric, the oils in the leather are removed and the leather becomes dry and brittle, cracking and tearing easily.

An accumulation of oil on the edge of a collar may be removed by careful hand rubbing with sawdust soaked in a suitable cleaning fluid. However, this is no substitute for the necessary over-all yearly cleaning most furs require.

Glazing: Glazing is the method used to restore luster to a fur. To the furrier, this formerly meant a careful wetting down of the hair in its natural direction,

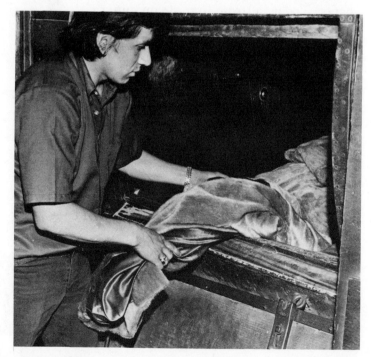

Cleaning process: fur drum contains sawdust which has been soaked with cleaning solution. Furs tumbled in drum retain natural oils.

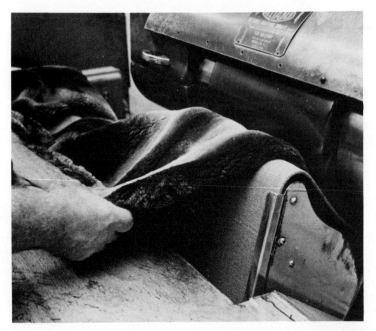

Rejuvenating furs is important. Machine at left freshens and firms up the hair. Below, steam gun is used to glaze a mink coat.

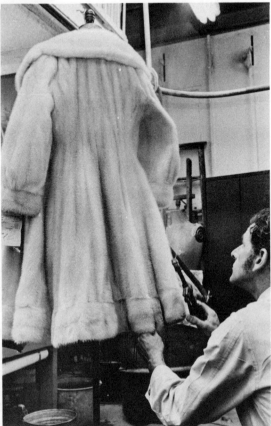

followed by reversing the hair using a dull-edged stick. Sheared furs were laboriously hand ironed with a carefully controlled iron wrapped in dull, unglazed paper. Today, a steam iron, steam gun, or revolving heated iron unit is used, depending on the fur. The appearance of curled or wavy-haired lamb garments is now improved by heat pressing on a standard dry cleaner's processing buck, with the steam disconnected. This process is usually a part of the fur cleaning. Some furs, such as black persian lamb, have this sheen added by spraying the fur with a lusterizing spray. Flat curled furs, such as the Russian broadtail, have wax paper pressed to the fur to give it the flat shiny look.

Bleaching: With the increased popularity of light-colored or naturally white furs, the problem of maintaining whiteness has increased. Bleaching, which involves the use of chemical treatment on furs, requires an extensive practical experience in working with fur rather than a great knowledge of chemistry. Fortunately, most furs in the light or white shades can be improved by this procedure.

Minor Repairs:

Fabric Tear: This usually occurs under the arm or in back of the armhole seam when too much strain has been exerted in reaching or lifting with fasteners closed. The pocketing will also wear away or tear when repeatedly subjected to overstuffing, or by the sharp edges of keys. In both instances, a temporary repair can be made from the outside, but a professional repair requires opening the lining edge nearest the tear and re-sewing from the inside. The lining is then re-sewn. Consideration should be given to the age of the coat and the condition of the leather. If the coat is old and the leather dried out tears will be common and repairs temporary at best.

Hook and Ring Fasteners: The matching crochet hook and ring, after constant wear and/or abuse due to excessive strain, will either be torn away from the fur or unravel, baring the metal ring which was originally covered by the crocheting. The hooks, as a rule, are more firmly anchored into the fur, and therefore less likely to be torn out. They do, however, become loose and may break

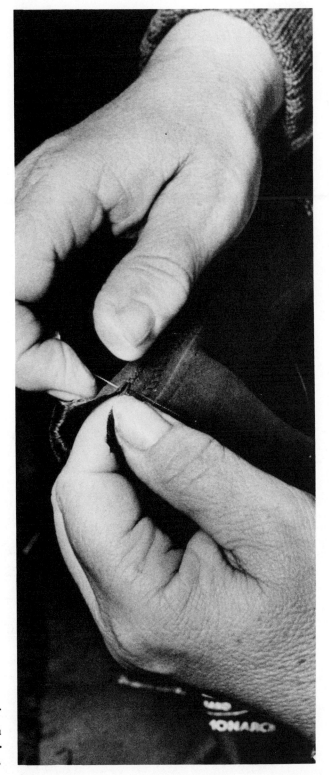

Sewing a tear in the leather of a fur garment. Tears can also be repaired from the fur side.

off. The crochet ring is sewn to the fur leather from the top, and can be replaced by a talented non-professional, but the hook requires professional expertise.

Small Tear or Open Seam: The fur sewing machine uses a single-thread stitch which, under certain conditions, can become unravelled. At other times the opening is caused by a tear. The cause of the opening should be analyzed. The difference between a seam that has unravelled and one that has torn is easily recognized under close inspection, even by the layman. The former shows an even edge and perforations, while the latter has the perforations torn through. The repair of a seam that has unravelled is a fairly simple job. In a tear, however, not only must the opening be repaired, but an analysis must be made of the causes for this tear.

Is the garment too tight in this particular area? Is the leather dried out and brittle? Did the seam open up or tear because of strain? The cause of the tear must be eliminated if the repair job is to be of lasting value.

Pocket Edge Tear: This repair differs from the pocketing tear previously described because it involves a separation of fur from cloth or fur from fur, usually at the corners. In most instances this requires, in addition to resewing from the leather side, some provision for reinforcing the corners with canvas or tape sewn into place.

Bald Spot: Some flat, stiff-haired furs do not wear too well, and bald or worn spots occur. Usually, even those spots as small as a penny require a major technical adjustment. Only rarely can the edges

Fur garment fasteners often need repair. The crochet ring, shown at left, is one such example.

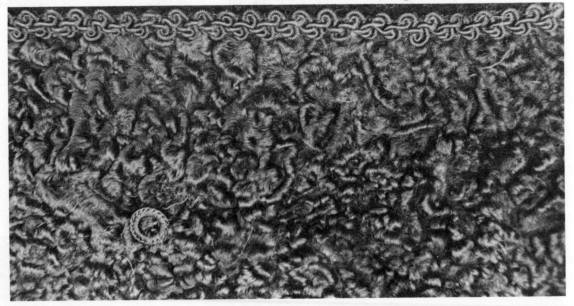

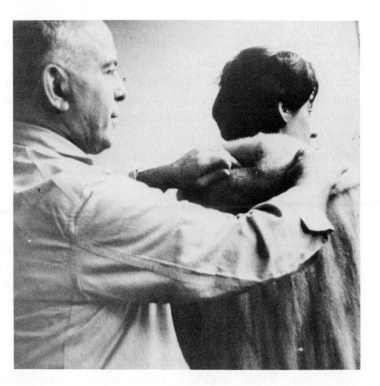

Remodelling represents an important element in the fur repair business. A new style may prompt a customer to change her collar, seen here.

of the fur be drawn together without creating a fold, gather or pucker. Thus, this cannot be considered a minor repair.

Minor Adjustments:

Until 1970 most length problems were confined to shortenings. Bulky furs do not lend themselves readily to shortening by doubling the hem. A lamb usually looks too bulky with more than two inches doubled or bent up, unless it is the thinner broadtail. If the fur is to be shortened more than this amount, it must be cut. It then becomes very difficult to lengthen, because of the problem of matching in the needed fur at some future date. A possible solution to this is to save the fur that has been cut off. Even here, a problem presents itself. The saved piece may not match the main garment when the customer wishes to add it because the garment may have oxidized to a different shade.

Resetting Fastener Positions: Fur coat wearers may gain or lose weight. Sometimes a replacement of the eyes in hook-and-eye closings or moving a button may give some freedom if a coat becomes snug. This will not, however, ease the increased tightness across the back and under the arms. Most of the time relief is possible only if material can be inserted across the back or under the arms. The former is often artistically impossible on highly patterned furs. The latter is a job for a professional.

Buttonholes: When buttons and buttonholes are used on fur, the buttonhole edge is protected from wear by grosgrain, leather, or similar edging. As this wears, it can be replaced. The difficulty which arises is not in the actual replacement of the material, but rather because of the necessity of completely opening the facing to get to the leather side to make the substitution, and the final re-closing of the front. These operations require a professional. The fronts must be resewn so that the two layers align perfectly and lie flat and even.

Major Repairs:

If a sizeable part of the area of the fur garment must be replaced by matching fur, the complexity and cost of the work rises sharply. The most common areas affected by normal wear are the bosom (from friction), cuffs (from jewelry), elbows and undersleeve (from driving a car), seat (from rubbing), collar edge (from hair conditioners), and front and pocket edges.

The immediate problem is to obtain matching fur which will not show when put in. This is sometimes impossible. The garment has undergone a certain amount of dulling and fading from exposure to light, air, and pollutants. A fresh skin inserted into the area as a patch will stand out like a sore thumb. The furrier must either stock or buy from a second-hand fur-matching dealer, fur of the same dye and color as the garment.

In large fur centers, well organized establishments specializing in used furs are found. Furriers who are not in these areas must mail in the garment, specify how much they need, and hope to receive a well-matched replacement section. In some instances it is impossible to find an exact matching replacement. Then a section from a less conspicuous part of the garment (the undersleeves) may be placed into the area needing replacement. An additional section can be used to fill in the less conspicuous empty area. The search for a good match may consume an inordinate amount of time. Some furs, such as Alaska seal and the spotted cats, are almost impossible to patch.

When a cuff or collar must be partially or completely replaced, an added complication ensues. Often the furrier repairing the garment is not the same one who originally made the coat. Therefore, he does not have the pattern for the unit. Even if the original furrier is making the repair, he may have long since discarded the pattern. It then becomes necessary to carefully flatten the work unit and strike off a pattern which can be used to remake the unit. The more worn the cuff or collar, the more difficult it is to make an accurate pattern simulation.

The most common major repair is the "complete re-edging" of a garment on which all major edges have worn away. This involves fronts, pockets, and cuffs. In most instances, the only way to repair the worn area is to cut it away and then rejoin the remaining edges. The average customer is aghast when told how much the work will cost and immediately asks why the removed fur cannot be replaced. This is an impossible matching job and cannot be done without the patch showing. The maximum width of worn edge that can be cut away on a side of the front edge is ½ in. on each side, or about 1 in. total. This means that front and facing will be narrower by ½ in. and the overlap will be diminished by the same amount. The other dimensions, however, will not be diminished. Collars and cuffs can be cut down as much as the worn edges indicate and will become correspondingly narrower.

The practical application of this technique is sharply limited when the fur is vertically striped or is sharply patterned, as for example, with the large cats. The amount that can be trimmed depends upon the effect on the stripe or figured pattern. Collars and cuffs may have to be removed and replaced entirely in this group of furs if they are badly worn. It often is more practical to introduce another fur to brighten the garment.

When a garment is badly worn, has already had an edging or two, or tends to wear rapidly, other techniques can be adopted to extend the garment's usefulness. The life of an older garment can be extended by two or three years by adding suitable decorative binding sewn over the worn edges by hand. This technique, however, is a last reprieve for the garment. The next step is to cut it down or remodel it. Nowadays, many garments are originally made with cloth, ribbon, grosgrain, or leather edging to minimize wear at vulnerable edges.

Lengthening a Fur Coat:

Some garments do lend themselves to a simple lengthening which extends the existing body silhouette acceptably. Quite often, the body line of a fur coat won't lend itself to just a lengthening, and must be completely remodelled. As we have seen, addition of fur so that the joining will be invisible is nearly impossible, except on black persian lamb and one or two patternless furs.

A far simpler and possibly more interesting method of lengthening a coat is to add a border. This can be in the same fur, in the opposite direction, or in a contrasting fur, which enhances the garment's appearance. It is sometimes possible to add the border by a hidden zipper, thus giving the owner a two-way adjustable length garment. Some of the more "radical" designers have even added non-fur borders, usually of leather or a fringed material.

Remodelling:

Suppose you have a one-car garage which is too small. You call in a builder and tell him that you want to enlarge it. He quotes you a price which is almost the same as your neighbor paid for a completely new garage. You protest, "Why so much? Look at all the material I have already." The builder properly points out that (a) he has to make a new plan (pattern); (b) the old garage must be dismantled; (c) not all of the old material is usable or will fit into the new plan; (d) the job will need full construction and a paint job. "I might as well buy a completely new garage," you complain. The comparative costs may bear you out.

The same applies to a fur remodelling project. A new pattern must be made and fitted, the old coat taken apart, worn or unusable parts must be discarded, the usable fur must be completely reworked to fit the new pattern, and additional fur must often be matched. Also, a new lining must be made and fitted to the garment because it doesn't pay to re-cut and re-use the old lining. Whether or not to remodel requires an evaluation of several factors, some of which are analyzed below.

To Remodel or Not to Remodel?

There is no fixed answer to this question. It depends on so many factors that each garment becomes an individual problem. What considerations should influence the customer and furrier in deciding whether or not to remodel? Which new style should be selected?

(a) **Age of the Garment:** It would be convenient to set down a number representing the age limit beyond which remodelling is not warranted. However, age is often secondary to the type of fur, the quality of processing, the amount of wear, and the style desired.

By and large, black persian lamb is the fur most commonly remodelled because it lends itself to such reworking. However, after 10 to 15 years, this fur may begin to *peel*, that is, the top layer of the leather to which the hair is attached begins to separate from the other layers in certain spots, exposing a spongy underlayer of leather. This is a sign that the leather has lost its resilience and the oils have begun to dry out.

In general, the popular-priced furs do not warrant remodelling because a high percent of their original cost was labor. This pushes the cost of remodelling too high proportionate to the value of the fur. In the more expensive garments, the investment in fur material is proportion-

ately higher and therefore remodelling may warrant consideration.

(b) **Type of Fur:** A coat whose original cost was $2000 may warrant a $500 remodel, while one whose original investment was from $200 to $300 may not warrant $75 for the remodelling. However, not all of the better furs lend themselves to every type of remodel. The spotted furs and Alaska seal are very difficult to patch, and these limitations apply to remodelling as well. Some very expensive furs like Russian broadtail are not too strong, thus limiting their ability to be remodelled.

Let-out furs present special difficulties, especially if the proportions of each stripe must be altered to conform to the new dimensions. This may mean that every seam holding adjoining skins together must be re-opened and some of the let-out cuts opened and resewn, or additional cuts made. This phase of let-out remodelling is so exacting that only a few establishments will undertake it, and only on fine garments where the necessarily high cost is warranted.

(c) **Layout:** The patterned layouts common to horizontal furs such as fox, fitch, rabbit, lynx, ermine, mole, and muskrat, present the same sort of obstacle to a drastic remodel. Certain areas of the pattern cannot be enlarged significantly beyond their original measurements. Nor can the layout itself be significantly remanipulated to any extent. To clarify the remaining problems, each of the major steps required in a remodel will be examined and its limitations and possibilities indicated.

Pattern Selection:

The average customer seeking a re-model wants her garment updated. This might mean making a short 38-39 inch straight, narrow body garment into a 43-45 inch tight-fitted garment with full skirt or even a midi coat. Both of these changes present radical departures from the original style, involve additional material, and in general require a major invest-ment of up to 50% of the original cost of the garment. The pattern selected, there-fore, must be simple rather than elabo-rate, restrained rather than full and luxurious and, above all, adaptable from the old. It is often far more sensible to remodel the coat into a shorter finger-tip walking coat or into a jacket, cutting down the cost of the remodel considera-bly and putting it within the realm of practicality.

To select a remodel style, a customer is shown a sketch, a cloth canvas sample, or a fur garment. Only the latter two give her any idea of what she is ordering and how it will look on her. Even the canvas often gives a false impression of the gar-ment's bulkiness. That model with the narrow bodice and full skirt may not seem to require any more material than is avail-able in the old straight-line coat, but the style makes the transformation highly im-practical if not impossible.

If a costly canvas fitting for a coat-to-coat remodel is required, the customer must be made to understand that the style will represent a compromise between her desires and what is feasible. These limita-tions are reduced if a smaller garment is desired. In the final analysis, it may be wiser to make an old coat into a jacket which will enrich and vary the fur ward-robe, and apply the saving to the purchase of a new coat.

Fur-Lined Cloth Coats:

What if the fur is too old and weak for further use? It can be sold, for very little, to a used-fur dealer or it can be

Many modern women, like the one shown here, purchase cloth coats but seek glamour—via a fur lining.

given away. If it is still presentable, however, its better sections can be reworked into an inside lining of a cloth coat. This type of garment is useful to the wearer who is susceptible to cold weather or who lives in a cold area. A *word of warning*: It is not possible for anyone to take an ordinary cloth coat and replace its lining with fur. Too many failures in such attempts have convinced furriers and amateurs alike that it is best to turn the work over to a specialist in making fur-lined garments. Whole patterns and styles have been developed to make such garments that hang, feel, and fit as they should.

Merchandising and Handling Remodels:

When styles were less varied and costs lower, remodels were more common than they are now. Any furrier capable of making a garment could also remodel one. However, when there was a large volume of this work, fur remodel specialists were common. Many department stores used these remodel specialists on a co-op basis, with the store supplying the sales, fitting staff and the advertising.

To remodel a garment the lining is removed and the parts of the garment separated into the major components— sleeves, collar, cuffs, and body. The special procedures include cutting away all worn areas and placing the best sections into the areas which are most visible. Cutting is more generous, since little reliance can be placed on stretching of the old fur. A more open sewing stitch, less likely to cause the fur leather to tear, is used. All parts are carefully taped by hand or with friction tape. If the garment is to be recolored, it must be cut extra large and given to the colorist, dyer, bleacher, or blender while still flat, after nailing and before squaring. Recoloring causes most furs to shrink so that renailing is necessary. This process may also weaken the leather and, in long-hair furs, singe the hair tips.

Fur Fitting:

Just as with a new garment, a fur remodel requires a fur fitting before lining. The initial pattern selection and measurement allowances become important here because, even more than in a new garment, the fit must be loose rather than snug. This applies especially to those measurements involving areas subject to movement stress, such as across the back, armholes, and bust. Because of the weakened condition of the leather, these areas are most susceptible to tears. Most remodellers produce a garment at least one inch longer than ordered, so they won't be "caught short" if a customer changes her mind. No amount of care in coming to final agreement on all measurements, location of buttons, and the like, is too great if it can avoid subsequent adjustments. For this reason slit pockets are marked but not cut through and patch pockets are not tacked on until after the final fur fitting.

The final stages of the remodel, lining, and finishing are routine. If the fur garment is lined with a wool or cotton flannel underlining, the change in fit may throw the fastener location slightly off. Most experienced fitters allow for this when placing the fasteners.

The used lining and the unused parts of fur may be returned to the customer. The customer is often shocked at the small amount of fur left over from the remodel. She cannot readily visualize how much fur had to be discarded because it was worn or weakened. It is common for the furrier to add fur rather than to have some left over.

Delivery and payment for the remodel does not end the furrier's responsibility. The customer, unless otherwise informed, may regard her remodelled garment as "new" and expect the furrier to keep it in good repair and condition for several years afterwards. This, of course, is not always the case. Many customers accept a reasonable time limit on the use of their furs if it is explained that the fur appearance has been improved but the fur itself has not been rejuvenated. Because of this limitation and the high cost, there has been a decline in fur remodelling.

Fur Storage:

Long ago handlers and wearers of furs found that the material was highly susceptible to mold, mildew, and moth damage when improperly stored. A certain temperature and humidity must be maintained when fur isn't being worn during warmer weather. Fur storage vaults work best when kept at about 40 degrees Fahrenheit, with a relative humidity between 45 and 55 percent.[1] Of course, fur storage also serves to protect against

theft. It may be a minor nuisance to the owner, but fur storage preserves the value of her cherished possession.

An approved fur storage vault is expensive. The furrier must use a storage service that has passed a rigid inspection by an insurance company inspector. The latter is far more knowledgeable and critical than any furrier or customer. The vast majority of furriers with average storage volume do not find it economical to build their own fur storage vault. Instead, they rent space in a certified, approved fur storage vault and store their customers' garments there. This procedure is so common that some of the larger department or specialty shops that have large fur storage volume use such outside fur storage warehouse facilities. The customer is helped rather than shortchanged. By hiring only as much fur storage warehouse space as needed, the cost of the service is kept at the lowest possible price.

The other major factor in determining the cost of fur storage is insurance. The customer may have a policy which protects her while she has possession of the fur, but once the fur storage firm issues a receipt and takes temporary control, it is liable for any damage or loss occurring through his negligence. The garment may also require other servicing and processing, which increases the number of hands and locations through which it must pass and thus also increase the possibility of damage or loss.

The competition for the fur storage business is so keen and the fees charged so minimal that most furriers are happy to break even on this service. Why, then, bother with it at all? It is a service which, while not profitable by itself, hopefully leads to or protects other profit potentials. Merchandisers refer to such sales or services as "loss-leaders." The furrier gives the service because he wants to keep his customer dependent on him for all of his fur needs. In addition, he counts on a percentage of the stored coats requiring some additional service from cleaning to remodelling on which he can make a legitimate profit.

Whether or not the furrier or store which offers this service has its own vault, they usually prefer to have a day or two notice when the customer wants his garment. This enables them to take it out of storage, and in most cases, blow it out a little or otherwise freshen it up before returning it. The garment has been cold and dry for some months and needs attention to bring the hair back to its normal appearance. If the fur is stored off premises the average notification required is three days.

FOOTNOTE—CHAPTER 7

1 "Fur Storage, Fumigation and Cleaning," *National Fire Protection Assoc.*, Boston, 1969.

8

Imagination Unlimited:
Unconventional Uses

8. Imagination Unlimited: Unconventional Fur Uses

How was fur first used? Was it utilized as a floor covering, a wall hanging, or a shelter? The answers, while of archeological interest, are not important for our purposes. What *is* significant is the modern resurgence of fur as floor and wall decorations, bedspreads, upholstery material, pillow covers, decorative throws over furniture, shoe fabric, and the like. These uses of fur are not new. They have roots in times past when fur was often the best available material. Even today, in some primitive areas, people live in homes made from animal skins and use furs for the same purposes as our ancestors.

There is a continual search for different ways to use fur in interior decoration, wearing apparel, and other areas. A polar bear rug before an open fire was for a long time the epitome of the romantic setting. Currently, however, the rarity of polar bear skins plus the cost of mounting add up to an expensive cost of approximately $2000. Modern decorators, striving for similar effects without exorbitant cost, have substituted steer, sheep, and zebra hides for polar bears. The development of the multi-colored pieced fur plates in elaborate designs has provided another source of decorative material and of apparel ornamentation.

Even these comparatively simple uses of fur present problems to the creative but untrained experimenter. If a sample made in a foreign country has been investigated and graded, all too often the bulk of the order arrives in dimensions and quality far below the sample. The items tend to dry up if constantly kept indoors in a heated room. The leather becomes dry, stiff, and cracks easily. The hair or fiber loosens and falls out as the leather dries up.

Nevertheless, with these limitations in mind, the opportunities are endless for ingenious uses of new forms of the old furs plus original combinations. Time and time again an entrepreneur not directly connected with the fur industry has seen one of the new furs in skin or plate form and thought of a new way to utilize it. This requires the type of individual who will seek out the sources of such furs and devise variations and applications.

For several years in succession, the millinery business volume has been largely concentrated in fur hats. Utilizing expensive and fashionable imported wooden hat blocks which are changed every year, the fur-hat manufacturer has captured a large part of the market by combining fashion appeal and comfort in a wide range of styles including caps, turbans, cloches, tams, and berets. Fur hats have been revived as a fashion for men as well as women. Their warmth, comfort, and attractiveness have practically insured their enduring popularity.

Fur coats for men have also enjoyed popularity in recent years. The coats have

FURS USED IN SOME UNEXPECTED FASHIONS

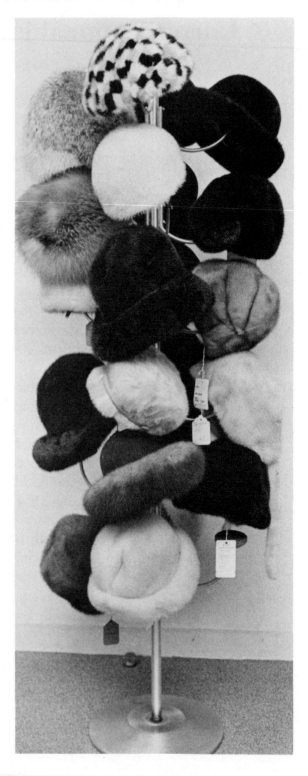

Furs are taking many shapes and forms. And the end is not yet in sight. Top, a variety of fur cushions. Below, furs used for a high-priced doll. At right, a hat tree laden with furs, mostly in mink.

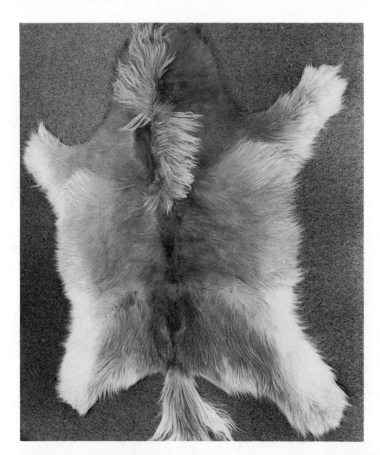

Two more ways to show
furs to advantage:
Left, the fur rug;
below, a variety of
fur coats shown in
children's department.

ILLUSION: ANOTHER FASHION WEAPON

Development of the silk screen technique to create the illusion of a particular fur is shown here. Below and at right, leopard design is produced on non-leopard skin. Bottom, a collection of stencilled furs.

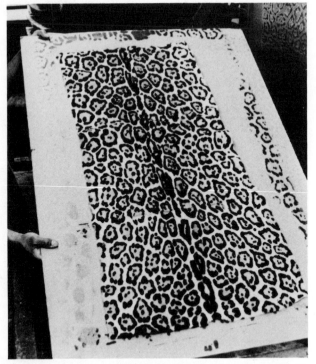

FURS CAN SHOW UP ALMOST ANYWHERE

With imagination, fur designers have worked furs into almost every type of garment. A few examples: Upper left, as lining and collar on cloth coat. Left, as persian lamb vest. Above, as shawl collar for sweater.

Fur retailers some-times use the product for decorative pur-poses. Above, as chair cover. Right, skins create wall display.

the fur to the outside, not as a lining, and are made in all colors rather than exclusively dark-brown and black. In addition to the more conservative furs, men's coats are being made in highly patterned, brightly colored furs. This is a revival of the once prevalent men's fashion of wearing brightly colored elaborate outerwear.

Pillows, wall hangings, and even bed covers of furs are currently being advertised and sold, even though the physical need for items of this nature in fur has disappeared in all but the most primitive cultures. A Westerner who lived with Eskimos wrote that "Caribou (fiber) hairs make fine Eskimo bedding and clothing . . . (because of) its hollow hairs which insulate far better than any fabric."[1] Here again "new" uses for furs will come into fashion to widen their use.

FOOTNOTE—CHAPTER 8

1 Guy Mary-Rousslière, "I Live With The Eskimos," *National Geographic*, Feb. 1971, p. 189.

9

Ecological Battle: The Impact of Conservation

9. Ecological Battle: The Impact of Conservation

During the 1960's, conservation became a most heated cause in the United States, and some of its most fervent supporters lobbied for programs that could have seriously affected the survival of the fur industry.

In general, the argument grew that Man was bent on destroying nature's creations: plantlife and animals. Also, Man used cruel instruments in his destructive pursuits.

When these ecological arguments were raised, the question of killing animals for their pelts inevitably arose. During the 'sixties and early 'seventies, the issue was transformed into a Cause which became a rallying cry for more than one organization.

Possibly the two most prominent groups in what came to be known as the "anti-fur" campaign were the Friends of Animals, headed by Alice Herrington, and The Fund for Animals, headed by Author-Columnist Cleveland Amory.

Miss Herrington's organization formed in 1957 "to promote interest in and knowledge of the humane treatment and proper care of cats and dogs and to carry out all activities connected therewith." In 1965, she amended her purpose to include the "proper care of animals." Her group was working for a "ban on shooting, trapping, poisoning or any form of slaughter of any and all wildlife on Federal public lands. In instances of proved over-population of any species, a corps of rangers, tested for sharpshooting ability, shall be designated to kill the old and weak of that species, only in instances where it is impracticable to relocate the excess or to reintroduce predatory animals."[1]

For his part, Mr. Amory issued a list of rights for animals together with a bumper sticker that read: "Animals have rights, too." One such right is "freedom from fear."[2] A joint effort by the pro-animal groups, led by Miss Herrington, reached a high point of emotional impact with a nationally distributed ad showing a mother and baby harp seal.

The headline over the ad read: "Battered Babies. There *Must* Be A Law" and the copy began: "This baby seal was lucky. It was dead before it was skinned." The copy continued: "There must be a law to end the slaughter of hundreds of thousands of baby seals every spring."

As the anti-fur campaign developed momentum, hundreds of thousands of dollars were raised for advertising and legislative lobbying purposes. New groups, or offshoots of existing protectionist groups, started to arise. For example, a group based in New York and San Francisco, called the Friends of the Earth, ran a full-page ad in the business paper, *Women's Wear Daily*, addressing an "important memo to the Fur Industry." The

The Snow Leopard

This year, next year—surely within this decade—one of the most magnificent creatures that ever lived, the Snow Leopard, will be gone from the earth forever.

Extinct.

Obliterated by the one force in the universe with the will and means to do it: Man.

Why do we exterminate him with such relentless ferocity? Did he threaten our lands, attack our children, diminish our food supply?

No.

40,000,000 B.C.- 197?

is simple and obvious:

Stop buying fur coats.

Your dollars support the slaughter. The lack of them will stop it.

You must believe that the wearing of fur coats is not a mark of affluence, but a symbol of indifference, selfishness and murder.

You must believe that as we kill all the wildlife, we doom ourselves to starvation. Life is dependent on other

Since fake fur is becoming all the rage—for men and women—when you do buy a fake fur coat, we hope you'll choose one made of our Timme-Tation.™

But our interest in animal conservation isn't entirely related to the sale of fake fur. Like you, we have families. Issues like Pollution, Conservation and Ecology are of vital, personal concern to us too.

Knowing what we know, we'd be fools not to try to save the animals.

And so would you.

Fur coats shouldn't be made of fur.

E. F. TIMME & SON, INC.

E. F. Timme & Son is one of the world's leading suppliers of plush and flat fabrics for Home, Industry, Transportation. 200 Madison Avenue, New York City.

WE WILL NO LONGER BUY *ANYTHING* MADE FROM THE SKINS, FURS, OR FEATHERS OF WILD OR ENDANGERED ANIMALS.

B

OUR GOAL is to make it unfashionable to wear wild animal skins. After only six months of organizing, we have had notable success. The names below represent only a tiny fraction of the total list of women *and* men who have subscribed to our position and who are spending at least a portion of their time encouraging others to join them.

Our purpose is *not* to drive an industry out of business. It is just that wildlife is vital to the stability of the world ecosystem—and no industry has a right to decimate it. When wild animal furs are no longer fashionable, something else will be and it will be up to your industry to anticipate the trend and *move* on it. Most of us do not object to furs which come from animals raised for that purpose.

This ad sponsored by Friends of the Earth, 451 Pacific Avenue, San Francisco; 30 East 42nd Street, New York, and the following men and women who will no longer buy products made from the skin, fur, or feathers of wild or endangered animals:

Mrs. Robert Alda
Mrs. Amyas Ames
Mrs. Didi Auchincloss
Mr. and Mrs. Emanuel Azenberg
Miss Lauren Bacall
Mrs. George Backer
Mrs. William A. Becker III
Mrs. Christina Bellin
Richard Benjamin
Miss Berinthia Berenson
Mrs. Leonard Bernstein
Mr. & Mrs. William Buckley
Mrs. Carter Burden
Robin Butler
Herb Caen
Cass Canfield, Jr.
Truman Capote
Mrs. Johnny Carson
Peggy Cass
Mrs. Jacob Cath
Dick Cavett
James Coco
Betty Comden
Mr. & Mrs. Barnaby Conrad
Mrs. Thomas Cushing

Blythe Danner
Ossie Davis
Dave DeBusschere
Ruby Dee
Mrs. Bruce Potter Dohrmann
Mrs. Roger Donaghue
Mrs. Garrettson Dulin
Mr. & Mrs. Jules Feiffer
Mr. Joseph Fox
Mr. & Mrs. Ben Gazzara
Mr. & Mrs. Frank Gilroy
Mr. & Mrs. William Goldman
Miss Tammy Grimes
Mrs. John Gruen
Mary Rodgers Guettel
Mr. & Mrs. William Hamm
Huntington Hartford
Mrs. Ernest Hemingway
Dustin Hoffman
Mr. & Mrs. Hal Holbrook
Bill Hosket
Mrs. Jacob Javits
Dorothy Jeakins
Mrs. Richard Kaplan
Danny Kaye

Mrs. Clark Kerr
Mrs. Alan King
Mrs. Arthur Kopit
Mr. & Mrs. A. Hunter Land
Marget Larsen
Linda Lavin
Mrs. Pat Lawford
Mrs. Christopher Lehman-Haupt
Mr. & Mrs. John V. Lindsay
Mrs. Joshua Logan
Mrs. Thomas C. Lynch
Ali McGraw
Mrs. Norman Mailer
Mr. & Mrs. David Merrick
Mrs. Stuart Millar
Gary Moore
Mrs. C. Kilmer Myers
Gerald Nordland
Mrs. Patrick O'Neal
Mr. & Mrs. Jerry Orbach
Mrs. Fairfield Osborn
Margaret W. Owings
Eleanor Perry
Mrs. R. Stuyvesant Pierrepont
Mrs. Francis T. P. Plimpton

Mrs. George Plimpton
Don Porter
Mr. & Mrs. Harold Prince
Robert Redford
Doris Roberts
Marcia Rodd
Mrs. William Matson Roth
Janice Rule
Pete Seeger
George Segal
Mr. & Mrs. Neil Simon
Mrs. Arthur Stanton
Mrs. John Fell Stevenson
Mr. & Mrs. Peter Stone
Mrs. William Styron
Saint-Subber
Miss Anita Tiburzi
Marietta Tree
Mrs Gianni Uzielli
Mrs. William Vanden-Heuvel
Gwen Verdon
Betsy Von Furstenberg
Edward Winter
Miss Kristi Wittker
Mrs. James Wyeth

C

Against fur use in varying degree are: (A) Fake fur maker E.F. Timme; (B) Friends of the Earth; (C) Alice Herrington of the Friends of Animals.

ad began: "We will no longer buy anything made from the skins, furs or feathers of wild or endangered animals. When wild animal furs are no longer fashionable, something else will be and it will be up to your industry to anticipate the trend and move on it. Most of us," the ad continued, "do not object to furs which come from amimals raised for that purpose (i.e., ranch-bred) although, in fairness, many of the signatories object even to that—seeing it as a further example of humans placing themselves 'above' other living species—a position which is both ecologically untenable and morally unsound." Signing the ad were a host of notables from the theater, the media, the arts and public affairs including: Lauren Bacall, Mrs. Leonard Bernstein, Tammy Grimes, Dustin Hoffman, Danny Kaye and the John Lindsays.[3]

For the thousands who made their living in the fur industry, the bitterest pill of all was the "ecological" campaign by sources that stood to profit from a decline in fur sales, namely the so-called "fake fur" manufacturers. Most outspoken in this category was E.F. Timme & Son, Inc. whose anti-fur attacks were prominently displayed in the *New York Times Magazine*, *Life Magazine* and other nationally distributed publications. One such attack read as follows:

"Today, there aren't a thousand snow leopards alive in the world. If the fashion bloodbath continues, he will be joined in eternity by all other species of leopard, all variety of tiger, blue fox, red fox, white fox, cheetah, beaver, wild mink, wild chinchilla, vicuna, lynx, pin martin, black bear, brown bear, opossum, raccoon, squirrel, polar bear, jaguar, ocelot, magay, and dozens and dozens of other creatures cursed with beautiful hides. How do we save the animals? The answer is simple and obvious: Stop buying fur coats. Your dollars support the slaughter. The lack of them will stop it."

Some of the animals listed above were considered endangered, according to one or more of the international, national or state organizations responsible for keeping tabs on such facts. For example, the International Union for the Conservation of Nature and Natural Resources said in its *Red Data Book* that the following animals were in dangerously short supply: tiger, leopard, cheetah, margay, jaguar, ocelot, bear and vicuna. The U.S. Department of Interior in 1969 listed 23 endangered species, including those in the *Red Data Book*, but few others that meant very much to the fur industry, such as: Canadian kit fox, giant South American otter, red wolf, and Utah prairie dog. The Department of Commerce in 1972 included polar bear as well as Northern and Southern Sea Otter.[4] (For charts on endangered species see appendix A.)

None of these official listings of endangered species endangered the fur industry, but the industry was at bay in the face of the anti-fur crusade that threatened a boycott not only of wild furs which accounted for about 20 percent of volume, but of all furs, as indicated in the Friends of Earth campaign.

The American fur industry, through

its key organization, the Fur Information & Fashion Council (FIFC), developed a program with these ingredients:

1. The industry accepted the judgments of professional conservationists and naturalists on the question of endangered species and agreed not to sell these species.

2. It created the Fur Conservation Institute of America to counteract what it called emotionalism and misinformation by its opponents. For example, it met the so-called "baby seal" attack with the following:

"The baby seals you have seen in films, television and newsprint are the Canadian harp seals. Their pelts are sold in the European markets only, and used primarily for leather. The American fur industry looks to the Pribilof Islands near the coast of Alaska for its seal skins. Here the seal population, harvesting methods and quotas, are carefully monitored by the United States Government to permit the herds to thrive. The sixty-year-old Alaska Seal Program (with Canada, Japan and the Soviet Union) is the outstanding success story of management and conservation, hailed by such recognized conservation groups as the Audubon Society. In 1912 the seal population of the islands stood at 215,900; today (1972) it is more than six times that . . . the herd is maintained at a level which the islands can support. . . ."[5]

3. The fur industry sought to provide funding, which would indicate a sincere desire to support humane techniques in the animal catch. For example, the industry's conservation arm agreed to contribute funds to two Canadian universities that had formed a Humane Trap Development Committee. It also helped create the Foundation for Environmental Education which, in the early 'seventies, was making study grants to students and educators in colleges throughout the United States.

4. The industry met some of its critics head-on. It charged that many of its opponents were, in fact, "preservationists" who "would have nothing killed, not even a rat." Continuing, an industry pamphlet contended the preservationists "don't care if deer starve for lack of food or seal battle one another for lack of room on a tiny island on which both space and food are limited."[6]

The industry also struck back at the "fake fur" adversaries: "Synthetic pile fabrics designed to resemble furs . . . come from fibers whose sources are not renewable. Whether these fibers are made from natural gas, petroleum by-products, coal or other raw materials, these resources are disappearing. . . ."[7]

5. The industry tried one more tack. In effect, it declared itself an endangered species, running a series of dramatic trade ads showing pictures of sad-eyed furriers, with copy like this to match: "They're out to get you. You, a furrier for so many years. They have their sights on you. Not with a gun. With words and pictures that can destroy what it took all those years to build. Who are 'they'? They're like that nice man you see on TV. A conservationist. With white hair, maybe. Or maybe

ANIMALS CONSIDERED HIGHLY ENDANGERED

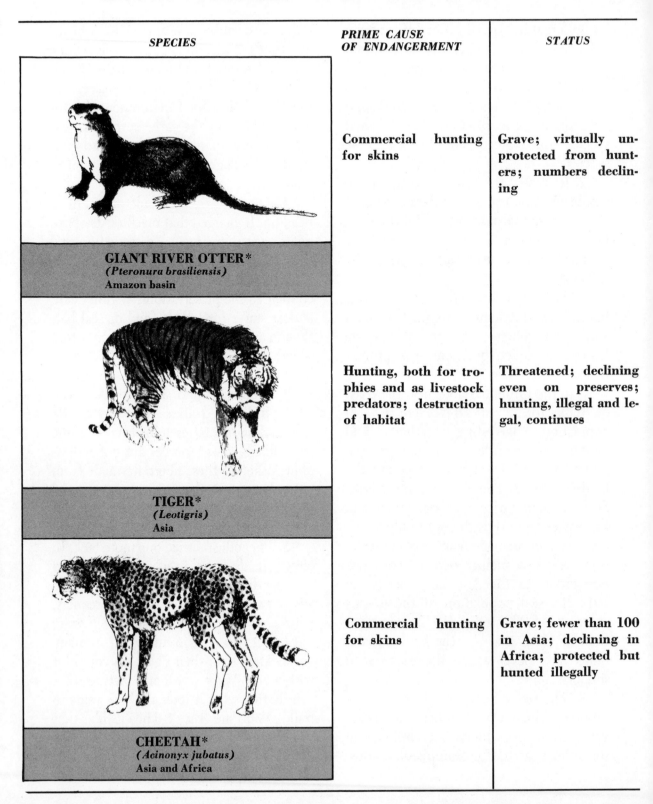

SPECIES	PRIME CAUSE OF ENDANGERMENT	STATUS
GIANT RIVER OTTER* (*Pteronura brasiliensis*) Amazon basin	Commercial hunting for skins	Grave; virtually unprotected from hunters; numbers declining
TIGER* (*Leotigris*) Asia	Hunting, both for trophies and as livestock predators; destruction of habitat	Threatened; declining even on preserves; hunting, illegal and legal, continues
CHEETAH* (*Acinonyx jubatus*) Asia and Africa	Commercial hunting for skins	Grave; fewer than 100 in Asia; declining in Africa; protected but hunted illegally

SPECIES	PRIME CAUSE OF ENDANGERMENT	STATUS
ORANG-UTAN* *(Pongo pygmaeus)* Sumatra, Borneo	Destruction of habitat; capture for zoos and labs	Grave; fewer than 5,000 remain; protected but habitat is still being destroyed
VICUNA *(Vicugna vicugna)* Andes Mountains	Commercial killing for wool	Precarious; fewer than 4,000 remain; some on preserves but vulnerable
TASMANIAN WOLF* *(Thylacinus cynocephalus)* Tasmania	Bountied because they killed sheep; habitat destruction	Precarious; protected but too few may exist for species to survive

* Protected under U.S. Endangered Species Law

just like that boy with long hair, no older than your son. They mean well. They want to keep the water pure and the air clean. And rare wild animals alive. But so do you, Ben, so they've got it all mixed up. They somehow believe you're responsible for the unregulated killing of rare wild animals for their fur. We're going to set the story straight."

At *Vogue Magazine*, the industry effort appeared to have borne fruit in the early 'seventies. The fashion magazine would be guided in its photography coverage by the list of endangered species provided by the U.S. Department of Interior. "It's important to remember that many species of wild animals are not endangered, not dwindling, but abundant, sometimes too abundant for the available habitat."[8]

But the Friends of Animals and its allies were scoring points, too. Heading the fur industry's Ecology Committee was George Stofsky, who reported that "state and city legislators are introducing bills banning furs other than those that are on the endangered species list of the Federal government. In California and Maryland, the purchase or sale of fur seal is now prohibited and in Cincinnati, the sale of all species of cats is banned. These laws seriously affect the entire fur industry and would indicate that eventually muskrats, raccoons, beavers, nutria and other abundant wildlife could easily be outlawed by legislators."[9]

From the industry's viewpoint, possibly its strongest asset lay in its identity with professional naturalists and conservationists, a highly respected breed during the ecological fervor of the 'sixties and early 'seventies. Outstanding in this group was Richard G. Van Gelder, Chairman of the Department of Mammalogy for the American Museum of Natural History in New York City.

Dr. Van Gelder reviewed some of the factors leading to the attack on the fur industry: Human population had quadrupled between 1850 and 1970. In 1850, the equivalent of 42 acres of land for food, shelter, clothing and minerals existed, on the average, for each person, but by 1970 that had shrunk to 10.5 acres. Actually, population growth increases about one million every five days, requiring that more and more land be taken from wildlife and converted into agriculture to feed people. Thus, the continuing squeeze resulted in the anti-fur crisis.

Dr. Van Gelder also responded to two of the essentially philosophical issues raised by the anti-fur groups: What right does man have to kill wild animals? And even if he has that right, must he do it?

On the right to kill, Dr. Van Gelder asserted that "there are no 'rights' in the natural world—to the victor belong the spoils. But there are natural laws, established over eons of evolutionary trial and error. These laws indicate that creatures who destroy their environment, foul their own nest, over-populate excessively, and kill too many of the animals on which they feed, carry with them the seeds of their own destruction. Biologically, then, man is another animal and he has equal rights to utilize the environment, but not to abuse it."[10]

On the argument against killing, the

naturalist contended that "the vast majority of animals that are killed are killed for food or to protect man's food. If no killing of any animal were undertaken, the immediate effect on man would be to increase disease and decrease food supply."

And on the issue that appeared in the 'seventies to draw the most support for the anti-fur groups, Dr. Van Gelder responded to the issue of clubbing baby seals to death: "If man may kill animals, the . . . least cruel methods are those that produce the least anguish and render the animal insensate in the shortest time. . . . A sharp powerful blow on the head, electrocution, and decapitation are, then, the most effective and fastest. None of these is pretty, and there has been much outcry demanding more cosmetic killing, rather than less cruel methods. . . ."

Somewhat paradoxically, in 1970, Dr. Van Gelder found himself in opposition to fur trade personnel when he testified for the so-called Mason Act, which created an endangered species list for New York State, protecting such animals as polar bears, leopards, cheetahs, jaguars and tigers. He was appalled at the furriers' complete ignorance about wild animal resources. "Now, when their livelihood seemed threatened, they lashed out blindly, mostly in the wrong direction, with little understanding of what was happening."

Dr. Van Gelder, however, found himself very much caught in the middle. He was "inundated" with literature from anti-fur animal lovers whose range of interests and emotions paralleled those of the Friends of Animals and similar groups:

"Some were concerned lest certain species become extinct; for others the problem was not the threat of extinction but the methods used to kill animals. Some sought my support in legislation, ostensibly well-intentioned, which, in practice, could turn the clock back a century to man's lowest relations with wildlife. Many of these people, like those in the fur trades, spoke ardently, but acted in ignorance. They knew so little of the situations of which they were so vocal."[11]

In charting his personal course, Dr. Van Gelder suggested that his answers "might seem inconsistent" to some people. For example, he opposed furriers who wanted to kill endangered cheetahs, but supported the fur industry in maintaining the Alaska fur seal catch. He objected to the quantity of harp seals being killed in the western Atlantic, but tweaked the emotionally upset who objected to the clubbing of fur seals. He asked "why they were not equally concerned that the same method is used to kill livestock that serves as our food."

The fur industry found another powerful ally, the hunting industry, without even trying, and it numbered in excess of 20 million. Sportsmen, through a host of lobbying groups and publications, concluded in the early 'seventies that the anti-hunting groups which, in this instance, were also the anti-fur groups, would not prove ephemeral.

"It would be nice if anti-hunting groups like the Friends of Animals and Fund for Animals and their outspoken supporters could be viewed as a passing fad of little importance," *Field & Stream*

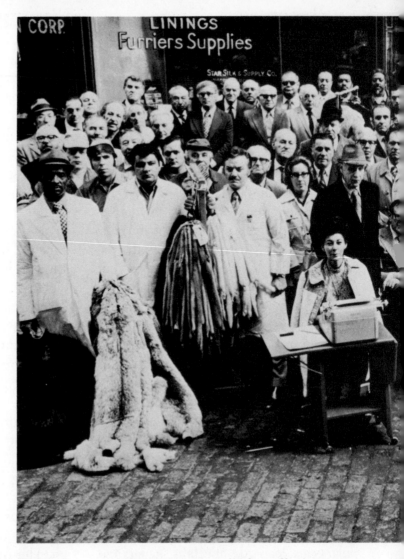

Industry strikes back. In dramatic effort to counter opponents, furriers pose here in heart of fur district. Now turn page . . .

noted in 1972. "But, unfortunately, this cannot be done if a concept of successful wildlife management is to be saved. These groups attack the traditional concept of hunting, and they will use any method they can dream up to achieve their goal. They do it in the name of humane treatment and environmental salvation, but the legacy of their victory could be the opposite of both. Such lessons have been given us in the past, and there is little wisdom in ignoring the lessons of history."

So, the appeal continued: let's launch a massive campaign. "After all, the membership of Friends of and Funds for Animals added together wouldn't reach 100,000, but there are some 20 million people who hunt. Your letters and your votes can make a difference."[12]

The attack against the anti-fur groups was even more pronounced in the 1972 *Shooting Times.* "Cleveland Amory: Go Soak Your Head!" screamed the article, addressing Mr. Amory the leader of *The Fund for Animals.*

"And the same advice applies to all

those 'preservationists' who substitute opinion for fact over hunting and wildlife conservation, issues far too serious for such folly. The anti-hunters are the most dangerous enemies wildlife has ever faced."[13]

The article made this case for hunting:

1. Hunters, by purchasing licenses, provided over $2 billion in wildlife preservation support.

2. Wildlife areas supported by this licensing was open to non-hunters and even anti-hunters, indicating that hunters were subsidizing the recreation of their enemies.

3. The "preservationists," by trying to abolish hunting and restrict shooting, are "eroding the prime source of funds for wildlife and wildlife refuge areas."

4. Hunting is supported by professional conservationists and government officials as "a key factor to wildlife management."

As the issue continued to boil in the 'seventies, the two most prominent anti-

fur groups indicated modest but recognizable differences on one aspect of the issue: trapping. Since the technique of killing was central to the emotionalism of the issue, the acceptability of trapping was crucial to furriers.

The Fund for Animals asserted that traps per se were evil. "Traps that hold animals were instruments of torture," a spokesman for that group stated.[14] Mr. Amory's group, however, limited its opposition to leg traps. In the early 'seventies, two types of traps were prevalent. The leg trap caught an animal attracted by nearby bait, and held the animal until it died from cold or starvation, or was otherwise killed by gunshots from the trapper. The other trap, called the Connibear trap after its inventor, Frank Connibear, was introduced in the late 'fifties and was considered by conservationists to be more humane because it worked like a mousetrap and caused almost instantaneous death. It had two drawbacks: it could not be safely set near domestic animals, and it was more expensive than the leg trap.

On the question of traps, Dr. Van Gelder noted that "most states have laws that require the trapper to visit traps at frequent intervals so that animals do not suffer unduly from the trapping or from exposure." Due to continuing concern by animal groups, and continuing attacks against fur trapping, the fur industry's conservation arm publicized the fact that it was contributing to various studies aimed at creating a "more humane" trap.

The ecological issues raised about furs were even more complex than Dr. Van Gelder, a student of nature and history, imagined. In addition to the arguments over humane animal treatment, the philosophical position of Man versus other animals, and the question of nature's balance and animal freedom, were considerations to changing fashion tastes, economics, cultural patterns and the existence, in the United States, of fifty different sets of state conservation laws.

On the cultural issue: In the early 'seventies Congress wrestled with a number of bills aimed at protecting marine mammals. While these bills differed in many respects, in principle they agreed on one point: the law should not restrict the Indians, Eskimos and Aleuts in the Pacific and Arctic Ocean coasts from killing these mammals because such killing was an adjunct to the culture of these people. In this instance, at least, culture became the dominant factor in making an ecological decision.

Also testifying to the complexity of the issue were African spokesmen who were not at all enthralled by U.S. decisions that placed leopards and other African game on American endangered species lists. A conservationist from the African nation, Tanzania, stated that "A law prohibiting the import of leopard skins to America would be a big blow to game preservation here because, in order to promote wildlife management, we must be able to show that game, through foreign exchange earnings, is an economic asset to the nation."[15]

In 1973 some Africans were fearful that the elimination of the U.S. market for their skins would result in animal over-population, and great damage in their African homeland. One African leader suggested that "the American conservationists and animal lovers should use their money to establish game preserves in the United States instead of letting animals"—presumably declared endangered species—"destroy coffee plantations, cattle and even children in Africa. We have plenty of animals and would be happy to supply them for such preserves."[16]

Pollution and the introduction of insecticides on a broad scale must also be included in the complex equation. There was suspicion among biologists in the 'seventies that the mating habits of such diverse species as the pelican and the elephant seal were being adversely affected by insecticides which had been washed off broccoli and celery plants. Similarly, there was consternation among the native Aleuts of Alaska when scientists reported that their favorite food, the liver of Alaska fur seals, had been contaminated; the livers had been found to contain 116 times the amount of toxic mercury considered safe for human consumption.[17] Oil, in one form or another, seemed to be adding its devastating potential to the ecological problem. In Northern Manitoba, Canada, the polar bear, considered an endangered species, was further beset by oil companies planning to explore areas previously reserved as the bear's reproduction bastion. Also, oil seepage off the coasts of Florida and California was fouling the water and smothering seals.

There was also the problem of the inadvertent destruction of a species, or of several species. Anxious to rout prairie dogs in the Black Hills area of South Dakota, the Government's Fish & Wildlife Service followed a policy in the 'sixties and 'seventies of using so-called "1080" poison and cyanide to cull the dog population. However, the head of the National Trappers Association complained that these poisons were also eliminating raccoons, ringtail cats, and bobcats, among other species.

There were times, too, when a fashion change produced a series of conditions endangering the safety of the human population. A case in point was the red fox, a known carrier of the rabies virus. Interest in wild fox slumped in the late 'sixties, partially due to an increase in fox breeding, and hunters and trappers no longer found it profitable to take the animal. So, the red fox flourished in Vermont in 1970 until there were too many of them. The food supply ran short. The animals were weakened and became more susceptible to rabies. The state, therefore, appealed to hunters to shoot these foxes on sight to inhibit the spread of the dread disease.[18]

The sovereignty of American statehood also interferes with the natural order of things. For example, in 1970, Arizona paid a bounty on the mountain lion. Six states near Arizona called him a varmint which meant he could be killed legally by anyone at any time, without any questions asked. However, in the Northeast, the mountain lion was classified as a game animal. This meant he could not be trapped, and it limited the time when he could be killed and the number a hunter could take. In Arkansas, Florida and New Hampshire, he was treated as if he were endangered, classified off limits for hunter and trapper.

Calling a species endangered does not, of course, automatically stop the flow of skins. In fact, those who wished to violate the ruling went to great lengths to profit from the scarcity. While the American fur industry itself supported the validity of leopard on the endangered list, compliance was not universal. An investigation by the East African Wildlife Society discovered enforcement loopholes wide enough for truckloads of leopard skins to run through: dishonest custom officials approved false export declarations; illegal shipments were slipped across borders to countries that held leopard skin sales to be legal; skins were mislabeled to evade restrictions; and poachers, of course, continued to flourish wherever the price was high enough to make it worthwhile.

The more ambitious suggestions on how to create world order in the world of furs ranged from tragic to sublime. A reciprocal exchange of endangered species had its lighter aspects. For instance, on Ceylon's protected list in 1970 were the following: rusty spotted cats, larises, pangolins, moose-deer, dugangs, delft ponies, leathery turtles, imperial pigeons, lorikeets and blacktail godwits."

On a more serious note, Dr. Van Gelder, looking towards the last quarter of the century, found hope in the fact that people, despite sharp differences, were aware of the situation. He thought several principles might lead to what he termed "any semblance of stability."

"First, protection must be afforded animals that are critically threatened with extinction. This must be done on all levels: international, national, local, individual and economic. We know that most species may quickly recover in number if given respite from killing in addition to the proper habitat. Unfortunately, some species are so depleted that they may be beyond the point of recovery. Not only must some be protected and their habitat guarded, but Man himself must maintain his own population at a level that will not further jeopardize wildlife by pre-empting his land for homes and agriculture. Similarly, commercial users of wildlife must also realize that animals are not an unlimited resource that can be taken without limit. They must understand the principles of agriculture and evolution to maintain a stock to produce a sustained harvest and not exceed it. This would assure the preservation of species while also assuring the availability of the materials for commerce. . . .

"Hope for the future? Yes, but not without sensible and dedicated efforts from all the people concerned. Not with mere blind generalities, but with specific activities based on wisdom, implemented with understanding, and intended for survival of both beast and man, will a better world of the future be realized."[19]

FOOTNOTES—CHAPTER 9

1 Margaret G. Nichols, "Alice In Disneyland," *Field & Stream*, May, 1972.

2 Nichols, *op. cit.*

3 *Women's Wear Daily*, June 2, 1970.

4 Fur Conservation Institute of America, bulletin, Sept. 1, 1973.

5 *Fur Naturally*, pamphlet of Fur Information & Fashion Council, 1972.

6 *Fur Naturally, op. cit.*

7 *Fur Naturally, op. cit.*

8 *Vogue Magazine*, editorial, September, 1972.

9 George Stofsky, "*Ecologically Speaking*," report from Fur Conservation Institute of America, 1973.

10 Richard G. Van Gelder, "*Animals & Mammals—Past/Present/Future*," 1972.

11 Van Gelder, *op. cit.*

12 Nichols, *Field & Stream*, May, 1972.

13 Dr. George V. Burger, "Anti-Hunters Threaten Wildlife," *Shooting Times*, February, 1972.

14 Harold Haber, "Record Prices of Pelts," *New York Times*, Nov. 22, 1973.

15 Sandy Parker, "Trader on Special Ban," *Women's Wear Daily*, May 5, 1971.

16 Sandy Parker, *op. cit.*

17 "Tokyo Mercury Level Is Found in Seals in Pacific," *New York Times*, Oct. 30, 1971.

18 "Rabid Foxes Rampant Locally," *Stowe, Vt., Reporter*, March 26, 1970.

19 Van Gelder, *op. cit.*

10

The Industry Faces
the Future:
Technology, Consolidation,
Organization, Marketing

10. The Industry Faces the Future: Technology, Consolidation, Organization, Marketing,

MAINTAINING THE FUR PELT SUPPLY

The constructive answers to the conservation problem lie in the development of new techniques to raise, maintain, and improve the supply and quality of the pelts. This may take the form of ranching or herding. The most extensive and successful example of fur ranching is with the mink. In the approximately half century of its existence, this industry, starting from the medium-brown wild animals originally captured, has successfully bred 29 major mink colors. These range from pure white to a brown so dark as to appear as black. Each of the major colors has shades and hues, which, at present, gives the customer a choice of at least a hundred different color variations.

It can be safely predicted that more cross-breeding, plus the accidental appearance of mutations, will constantly increase the number of color variations available. For example, a recent development in mink breeding has been the introduction of *Kojah*, a long-haired mink resembling the sable. This may lead to a new direction in mink and other ranching in which not only color but physically new and distinct characteristics are artificially bred.

Historically, mink ranching had its origin in the United States. This industry, thanks to the export of American ranch-mink breeder stock, has now spread to twenty-one countries in Asia, Europe, and South America. The United States, the Soviet Union, and Denmark are the leading ranched-mink producers. (The bulk of the wild mink comes from the United States, Canada, and the Soviet Union.) Recently, the volume of ranched mink exported by the United States has dramatically increased, in large measure because of the competitive advantage of a devalued dollar as well as its superior product. Each year, however, a smaller percentage of the world's total mink population grown on ranches has come from the United States. This is to be expected as other countries perfect and improve their ranching techniques.

The other two methods of production, herding and establishing preserves, have the advantages of being less expensive and conforming more closely to natural ecological conditions. However, whether it be the Everglades or the Great Swamp of northern New Jersey, sooner or later the pressures of expanding technology and population force reconsideration of their continued existence. The wild animal roaming untouched on a preserve may be physically perfect, but the reflection of this in his progeny is a matter of chance rather than control. So far, only the lamb family has been of importance to the present-day fur industry as a herding source on land.

Extraordinary expansion of the pieced mink industry has led to a host of new fur fashions. Here, three distinctive looks: Above, mink coat; Top right, sweater; Bottom right, jacket.

Heads and Tails: Virtually every part of the mink skin can be used for garments. At left, jacket made from mink heads. Above, coat made from mink tails. Paws and sides also used extensively.

137

Processing Pelts:

One of the most imaginative and progressive segments of the fur industry has been the various allied processors, including the dressers, dyers, shearers, and printers. They have never hesitated to experiment, venturing often into new, untried fields. They have developed so many original effects on standard and previously unused furs that, with the exception of mink, and one or two others, over 90% by volume of all furs are used commercially in a form different from the original. New colors, treatments, and finishes on both hair and fur side of any pelt may suddenly restore or even create a demand for some fur not previously used.

The *Lakoda* seal is a good example. Until the 1960's all attempts to prepare the female lakoda Alaskan seal to make it commercially useful had failed. Only by departing radically from traditional seal processing was a commercially useful product created. The untrained layman cannot distinguish this fur from the traditional Alaska fur seal made from the male bachelor.

Another good example of this innovative spirit is the silk-screening of fur. The rage for spotted furs initiated by leading fashion-conscious women a few years ago, has mushroomed to other exotically patterned furs. Led by the craze for the all-too-rare North African leopards, the demand widened to include "new" furs like tiger, zebra, jaguar, cheetah, and giraffe. Interest has returned to neglected furs like ocelot, spotted cat, and lynx. To supplement the lack of natural material and to supply a more moderately priced

Furriers try some new approaches: For instance, fox on suede (above). Opp. page, left, a ranch-raised nutria coat; top right, silver fox worked diagonally; bottom, natural nutria coat with lynx collar.

item, the trade created a processed and stenciled fur imitation. Today, the skins are silk-screened. Fortunately, there are skilled craftsmen capable of creating the illusion of leopard, zebra, and others. The clarity and sharpness of the artificial pattern exceeds that of the best natural skins. The colors hold very well, turning dull a bit sooner than the natural colors, but costing a fraction of the price of the original.

The artists work from a sample or a sketch and carefully draw the pattern on a clear plastic plate using opaque India ink. The receiving silk screen is made photographically sensitive by suitable chemical treatment and the design is transferred to the screen by exposure.

The durability of a garment made from these stenciled skins often depends on the fur base used. The most common are kidskin, calf, antelope, kangaroo, lamb, and rabbit. The fur is usually prepared by bleaching, dyeing, and shearing to make the ground color and fur match the fur being copied in color and height of hair. Obviously, such treatment does not add to the strength of the fur. On the other hand, many of the originals have a short and stiff hair which tends to wear away at the edges.

Experimental designs are not limited

Creating looks that attract. Right, striper dyes guard hairs. Opp. page (l-r), pin-striped mink; lamb and fox, stencilled.

to imitations of natural furs. They have also tried abstracts and radical modifications of animal patterns. A good example is "baby giraffe," a miniaturized version of the large spaced irregular boxes found on the adult.

Replacing Fur Craftsmen:

In recent years, the emphasis in our society on getting a higher education has made it more difficult to interest young people in the crafts. The fur garment manufacturing industry, along with the apparel trades in general, has had to cope with this trend. Traditionally its ranks have been filled by waves of immigrants.

Germans made up the bulk of the furriers in the era before 1917. After the turn of the century through World War I, the Eastern Europeans, mostly Russians, and usually Jews, along with a sprinkling of Kastorian Greeks, began to swell the ranks of the fur craftsmen. Through the depression and into the years after World War II, furriers were envied by other workers in different crafts for their high salaries.

In the last two decades the picture has changed considerably. In New York City, still the largest center for fur garment manufacturing, a leading fur manufacturer listing shows a total of 1184 firms in

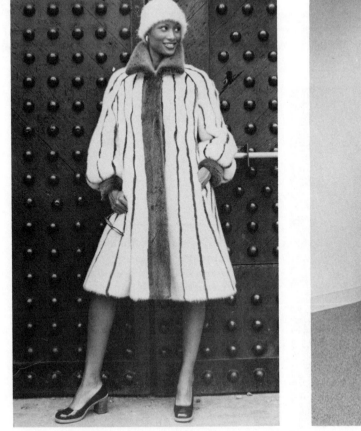

In the 1960's major efforts to get
more young people interested in
fur craftsmanship were undertaken.
Here, Instructor Ed Kleinman of
H.S. of Fashion Industries works
with career-minded students.

1970. In 1950, there were 2104.[1] The labor force has substantially declined from its post World War II height, and would be even lower were it not for a revival in the flow of Greek craftsmen. Frankfort, Kastoria, and Montreal have schools for furriers which attract many students. Israel is about to institute a fur training course to supply needed talent for their growing fur industry. In New York City, the Board of Education, through one of its specialized vocational high schools, the High School of Fashion Industries, has maintained a fur garment manufacturing department with both day and evening courses since 1926. This department has experienced in recent years, a tremendous increase in interest by both adults and youngsters in learning the crafts of the furrier. This welcome development has been encouraged by the fur industry through donations of money, raw materials for learning, and the placement of the students.

Fur garment manufacturing centers are springing up and growing fast in diverse areas of the world. Frankfort, Tokyo, Tel Aviv, Kastoria, Milan, Barcelona, Copenhagen, and Montreal are just a few of the fur centers which, though long in existence, have now begun to attain international importance. This is because of the development of a pool of skilled craftsmen, properly utilized by aggressive management. The traditional centers like New York, London, and Paris are holding their own, unable to expand in large measure because of the difficulty in attracting a still greater supply of highly skilled craftsmen.

This situation holds for manufacturing, but not for designing. Several well known designers have brought their cities to world attention through the impact of fur designs effectively executed and properly merchandised. Some examples are Birger-Christensen (Copenhagen), Jole Veneziani (Milan), Chombert (Paris), and Grosvenor (Montreal). Their creative success is cumulative within their organizations and in their geographical area, encouraging craftsmen, designers, and management to enter into or expand their efforts. Certainly, the world centers for fur have begun to shift.

The Status of the Fur Business in the Early '70's:

The individual craftsmanship which still remains a distinguishing characteristic of the fur garment industry is both a source of pride and an obstacle. Only one garment at a time can be cut, sewed, nailed or finished. Even today, the sewing of the skin of a thin leathered fur such as Russian broadtail is done by hand. The industry has failed to utilize all of the technological improvements available. An important factor inhibiting technological advances in the U.S. fur industry is the small size and great number of individual establishments on both the retail and wholesale levels. In New York City, the largest manufacturing firm produces less than one-fifth of one percent of the industry volume. Many small firms consist of two or three craftsmen, all of whom may be partners, with no one on the payroll as an employee. Such ventures are often marginally financed, so that they cannot afford to experiment with or invest in costly new and advanced equipment.

On the other hand, the possibility exists for one or two craftsmen who have saved up some capital to set up a small business for themselves. Many of the world-famous names in the fur industry started in just this way. Competition is varied and keen in contrast with other branches of the apparel industry, where business units have tended to become larger in size and fewer in number. To sum up, the fur industry is one of the last areas where the small entrepreneur flourishes.

Foreign firms have adhered more closely to the tenents of modern industry: capitalize fully, combine, develop, and make use of all the potentials of modern technology. Several individual firms are larger than any of their American counterparts, whether you count employees, capitalization, or production. Many have international impact. As a result, the increasing number of foreign-made fur garments sold on the United States market has become a matter of concern.

Fur Dealer:

Recent developments in fur skin marketing may lessen the economic influence of fur dealers and the relative importance of the auction houses in the United States and England. During the early part of the century, commission selling by private agreement was more important than it is now. The proposed resurgence of this technique of marketing may theoretically eliminate the fees and charges added to the cost of skins by the fur dealers and auction houses. But it leaves unanswered several major questions:

1. Is the commission house fee going to be less costly than that charged by the auction houses?

2. Who is going to pay for dressing the skins, or is the industry going to revert to raw skin selling only?

3. Who or what will replace the financing functions of the skin dealer?

Industry Organizations:

The answer to many of the above problems is obviously organization for the common good. The unionized fur workers are members of the Fur Department of the Amalgamated Meat Cutters and Butcher Workers of the U.S. and Canada. The fur union has locals in every fur manufacturing area where there are a number of workers. It is a craft rather than an industrial union. Each local is composed of the various crafts, such as cutters, operators, nailers, and finishers. Each craft has its own pay scale. Small groups are organized into a composite local.

Where there are hundreds of workers, the various locals combine for convenience into a geographical entity usually called a joint council. In New York City, for instance, the umbrella organization is the Furriers' Joint Council, a council of representatives of the various craft locals in the fur garment manufacturing center.

The corollary to the workers' union is an employees' association. The fur retailers have a national body known as the Master Furriers' Guild of America. The fur garment manufacturers are organized by localities, with the largest in New York

A lighter moment in labor negotiations, which are generally very tense. Here, union officials are in forefront.

City known as The Associated Fur Manufacturers, Inc. A smaller parallel organization, composed mainly of owners of Greek descent, is known as The United Fur Manufacturers. Many of the very small marginal firms are "independents" and do not belong to either organization. Some actually employ no workers at all and, therefore, have no contacts or contracts with the union. Any firm, however, wishing to employ a union fur worker must sign a contract with the union, usually similar to those in effect with the Association firms.

It would be a mistake to assume that the two economic units are always in opposition. In New York City over the years, they have found that by working together it is possible to bring greater

stability to the industry as a whole. To relieve the problems of the sick and the elderly worker they have jointly organized and funded a Health and Retirement Fund. The jointly supported Fur Label Authority sees to it that every fur garment is labelled so that the consumer can identify it as having been made under agreement conditions. All segments of the industry contribute to the Fur Information and Fashion Council. This agency, operating under a limited budget compared with other industry publicity institutes, has done much to promote furs to the public by means of shows, literature, and film.

Other countries, with smaller fur industries, have gone further to promote fur products. For example, France has its

Eye on marketing: Top, Adman Stanley Katz discusses campaign. Bottom, Ed Stanton, p.r. specialist, outlines fashion show.

"Comite d'expansion de la Fourrure." No such nationwide organization for fur garment manufacturers exists in the United States, except for the limited influence of the Fur Council of the National Retail Merchants Association.

In response to an obvious need, an attempt is being made to unite every segment of the industry throughout the United States in the newly formed Council of American Fur Organizations. The timing indicates that their priorities would be to counteract the onslaught of misinformed conservationists and to bolster the American industry against the growing competition from abroad.

Perhaps a weakness of the American industry is its multiplicity of specialized and segmented organizations. In New York City there is a separate organization for each of the following:

wholesalers (jobbers)
merchants (skin dealers)
resident buyers
blenders
brokers
dressers
American Legion Fur Post
an industry lodge of B'nai Brith
a charitable foundation (for the needy of the industry)

The ranchers, who have thus far shown a willingness to organize at national levels, are represented by EMBA Mink Breeders Association which works for those ranchers raising mutation colors, and the Great Lakes Mink Association representing the ranchers raising standard (dark-brown) mink. These two organizations have repeatedly made efforts to combine, and may yet find it possible to do so. Other ranchers here and abroad have their own associations through which they operate.

Saga represents much of the mink production of the Scandinavian countries. The Southwest African Karakul Ranchers Association promote their product worldwide under the Swakara label. In the United States breeders of such ranch furs as nutria and chinchilla have their own organizations. The U.S.S.R. fur peltry arm is Sojuzpushnina, which advertises and sells Russian furs. In Greece, an official "permanent Fur Committee" was formed in February 1971.[2] Obviously, some of these nationwide groups are backed wholly or partly by their respective governments. This situation places them in an advantageous position to finance equipment, production, publicity, and sales.

The Council of American Fur Organizations mentioned above offers the U.S. fur industry its first opportunity to present a united front to the public and to government agencies. The list of member organizations includes:

American Fur Brokers
American Merchants Association
Associated Fur Salesmen
Associated Fur Manufacturers
EMBA Mink Breeders Association
Fur Blenders Guild
Fur Dressers Bureau
Fur Garment Traveling Salesmen

Fur Wholesalers of America
Furriers Joint Council of New York
Great Lakes Mink Association
Hudson's Bay Company
Leather and Machine Workers Union
Master Furriers Guild of N.Y.
National Retail Merchants Association
Resident Fur Buyers Association
Scandinavian Fur Agency
Seattle Fur Exchange
United Fur Manufacturers

On a world-wide scale, the International Fur Trade Federation, to which 23 associations from 21 countries belong, has attempted to unite all interests in those countries into a world-wide body. The major fur organizations in the United States have not found it practical to cooperate, much less join. The 1970 meeting, held in London, had an agenda listing: a distribution of funds to member countries to promote wearing of furs; conservation; and suggestions to the Canadian and American auction houses on changes in auction procedure. It also strongly recommended a total ban on La Plata otter, giant otter, clouded leopard, tiger, and snow leopard and a temporary ban on leopard and cheetah. Cooperating with the International Union for the Conservation of Nature and Natural Resources, based in Morges, Switzerland, and headed by Dr. E. Budowski, they agreed to a joint investigation of the sustainable yields of other species of spotted cats.

FOOTNOTES—CHAPTER 10

1 Classified Fur Source Directory, *Fur Age Weekly*, 1970.
2 "Greek Fur Industry Sets Up Committee," *Fur Age Weekly*, Feb. 8, 1971, p. 1.

11

Profiles on All
the Important Furs

11. Profiles on All the Important Furs

Traditional treatises on furs have included extended alphabetized lists of the characteristics of each of the common furs. The trouble with the traditional charts or lists is the transient nature of much of the information given. Apart from the name of the fur, which is fixed by the Federal Trade Commission and the Department of Interior, nothing else is permanent. For instance, beaver should be classed as "wild," but experimental farms in Idaho and Canada suggest that it soon may be a "ranched" fur. The natural geographical origin may have no bearing on present sources, as witness "persian lamb" and mink. A fur which has always been used in one form may suddenly appear with some physical alteration, such as "plucked" mink, or natural marmot which has previously been used only in dyed form.

Furthermore, the durability of any given fur varies with its original nature, grade, whether taken at prime, the amount and types of processing it has received, the frequency of its wear and whether it has received proper care. Durability is no longer the prime factor in furs. The average customer no longer buys a fur desiring to or hoping to pass it on to her daughter. Fashion and tastes change so rapidly today that the average fur goes out of style long before it wears out. Costliness and durability correlate more closely in furs than in most apparel,

although some of the most expensive, such as broadtail persian and chinchilla, are not necessarily the most durable. In the following lists comparative durability is expressed in percentages derived from an authoritative chart worked out by a group of fur scientists.[1] The figures represent current findings and are obviously subject to change.

The information on use, repairability and remodelling is up-to-the-minute, and is also a variable. Marginal decisions on remodelling, for example, may be affected by the locality the customer is in, the time of the year she chooses to have work done, and so forth.

The wild U.S. catch figures are derived from the 1967–8 report. Preliminary reports on the 1968–9 season indicate a 14% increase in dollar value of the catch to $15,400,000, with price increases for all furs except marten, weasel and timber wolf. Decreases in the harvest of beaver, marten, otter, wolf, and wolverine were indicated. All other species report increases. Louisiana led in dollar volume with a catch valued at over $6 million.[2]

What furs should be listed? This poses serious questions about the extent, range, and sub-division of species. One list of seals compiled by a student of the field lists 93, many of them fur-bearing. The original list issued by the Federal Trade Commission in 1951 under the Fur Products Labelling Act contains 104 names. To show how confusing this can become, two of these were changed in an amendment in 1969. The list does not yet contain water buffalo (Zebu), zebra, or elephant, yet all three have since been used for fur garments. Many are as yet exotic and rare and too little is known about them as furs used for wearing. The following covers most of the better known furs, but it must be emphasized that ecological and ranching developments may make any item of information obsolete on short notice.

As of this writing, it can only be suggested that the following have been used as fur, or are being experimented with:

manul cat, desman, jackal, jaguaronoi, kangaroo cat, kinkajou, koala, jap marten, South American and water opossum, panda, Mexican raccoon, wombat, woodchuck, solongoi, sea otter, zebra, giraffe, buffalo and water buffalo. Not enough is known about them, or they have been so little used as fur that they do not warrant the thorough analysis given to other furs in the following pages.

For some, fur cross-references are necessary because of the multiplicity of names given to the animal. The following cross-reference list will assist in locating furs with dual nomenclature.

CROSS REFERENCES

Goatskin, *see* Kidskin
Cattle, *see* Calf
Genet, *see* Cat
American Ringtail, *see* Bassarisk
Rock Sable, *see* Bassarisk
Panther Cat, *see* Spotted Cat
Lynx Cat, *see* Wild Cat
Bob Cat, *see* Wild Cat
Chinchillone, *see* Chinchilla
Chinchilla Rat, *see* Chinchilla
Tiger Fitch, *see* Fitch
Perwitsky, *see* Fitch
Tanuki, *see* Cross Fox
Corsac Fox, *see* Kitt Fox
Moufflon, *see* Kidskin
Karakul, *see* Lamb Family

Kid-Caracul, *see* Kidskin
Peschaniki, *see* Marmot
Pine Marten, *see* Baum Marten
Kojah, *see* Mink
"Hudson Seal," *see* Muskrat
Ocelot, *see* Ocelot
Peludo, *see* Ocelot
Ferret-Badger, *see* Pahmi
Rock Seal, *see* Introd. Fur Seals
Tropical Seal, *see* Introd. Fur Seals
Cape of Good Hope Seal, *see* Alaska Seal
Zorrina, *see* Skunk
Russian Sand Weasel, *see* Susliki
Puma, Cougar, *see* Mountain Lion
"Gigas," *see* Mink

Explanation of Headings:

Size and Appearance: Applies to natural dressed dyed skin before trimming and cutting, except when working size is indicated.

Geographical Sources: Limited to major sources such as wild, herded, ranched, etc. Affected by ecology, conservation, etc.

Processing Used: Applicable only up to

time of writing. Will inevitably change.

Durability: Based upon scientific study, plus author's experience. See introduction to chapter.

Skin Layouts Used: Limited to trade methods generally used in last half-century.

Repairability: A variable, less limited in fur centers where labor is less expensive. New York market used as base.

Remodelling: Same variable as fur repairs, but basic limitations still apply.

Care: Self-explanatory. Also see appropriate questions and answers in Chapter XII.

Look Alikes: Can apply both to natural appearance of skin, or similarities after processing. Variants endless as processing improves.

Comments, History: Sharply limited suggestions selected from multitude of interesting facets and history of each fur.

Wild U.S. Catch 1967–8: Limited to U.S. wild furs, Official U.S. Department of Interior figures. Preliminary estimates indicate continued drop in volume in succeeding years.

Name: Antelope

Size and Appearance: Varies in size by continents. Usually colored light brown to gray to blend with environment. American varieties have white rumps. African types have short stiff hair, easily broken in friction. Fur in various shades of gray. African "Spring boc" has distinct white belly.

Geographical Sources: Northeast African—small brown pelt. Once used in trade, reintroduced in 1960. Australian reddish longer-haired, new to fur industry.

Processing Used: Dressed: Used in natural shades, or dyed in natural color is poor.

Durability: 10% to 5%.

Skin Layouts Used: Hair down, so far. Full skin layout.

Used For: Tried for natural coats in Europe in 1960's. Best manufactured with leather or other durable edging.

Repairability: Not advised, except for re-edging involving no additional fur.

Remodelling: Not advised.

Care: Edges wear quickly.

Look Alikes: Calf, Pony.

Comments; History: Periodic attempts may develop antelope skin as fur by improved processing, especially African subspecies.

Name: Badger

Size and Appearance: Size—medium dog. White line from tip of nose over center of head gives it its name. Color ranges from pale white to brownish tones. White grotzen varies in length and width. Almost as wide as it is long. Pale silvery best, yellowish tinge less valuable.

Geographical Sources: Wild. Europe, China and North America. Canada and U.S. (plains states). Canadian pale white, U.S. cream colored, Asiatic, yellowish brown.

Processing Used: Dressing: Generally used natural.

Durability: 80–70%.

Skin Layouts Used: Full and split skin. Can be let-out on limited basis. Leathered to give chevron effect.

Used For: Use varies greatly with fashion changes. Usually used as cloth coat, collar and cuff trim. Used occasionally on other furs. Now used as a fur coat with leathered chevron effect.

Repairability: Cannot be patched,

matching added fur almost impossible.

Remodelling: If in trimming, slight changes possible, if similar pattern. In garment form, not advised.

Care: Strong fur. Normal care and cleaning.

Look Alikes: Chinese and "Sand" Badger have yellower, coarser hair. Pahmi is smaller and has same head patch, brown to silver gray guard hair.

Comments; History: Heavy grotzen hairs used to make fine shaving brushes. Also used to make paint brushes.

Wild U.S. Catch 1967–8: 2,754. North Dakota and Nebraska leaders.

Badger

Name: Bassarisk: True zoological name for raccoons' cousin.

Size and Appearance: Tail and head resemble raccoon, brownish with yellow in color, size of large kitten, 16–17 inches long. Tail very long, grotzen, darker in shade.

Geographical Sources: North American exclusively. Almost entirely from Texas, few from Mexico.

Processing Used: Originally almost always bleached or dyed to imitate other furs. Now being used natural, in block horizontal layout.

Durability: 60–50% dyed, 70–60% natural.

Skin Layouts Used: Semi-let-out when dyed and used for trimmings. Block, horizontal and vertical for garments.

Used For: Originally a trimming fur, used dyed. In late 1960's occasionally made into blocked jackets and coats, dyed and natural.

Repairability: Not particularly difficult, so long as edges only involved.

Remodelling: If in block skin form, may be possible for slight pattern changes. Not let-out in U.S. Does not warrant labor costs if let-out remodel desired.

Care: Not for every-day wear, especially if dyed. If plucked, do not allow hair to become matted.

Look Alikes: Large dyed squirrel, if same shade. Both have soft underfur.

Comments; History: Much confusion on name. Incorrectly known as "American Ringtail," "Rock Sable," Raccoon, Fox, Cat, etc.

Wild U.S. Catch 1967–8: 23,364 almost all from Texas.

Name: Bear

Size and Appearance: Polar—largest fur-bearing animal, white, sometimes yellowish tinge, from Arctic regions. Black, North America. Fine leather, hair long, dense and silky, a deep black. Best from Alaska. Honey, brown and cinnamon are all phases of black bear. Grizzly—a silvery brown. Glacial and Kodiak are similar to grizzly in size. Good quality bear has hair 3–4" long.

Geographical Sources: Blacks, mostly from Canada. Grizzly from Rocky Mountains. Glacial and Kodiak from Alaska. Polar Bear from Arctic islands.

Processing Used: Polar and other bears used for rugs are "mounted," an expensive process which recreates natural appearance of head, paws, etc. This procedure costs several hundred dollars.

Durability: Polar, 100–90%, Brown 90–80%.

Skin Layouts: Solid skins or sections thereof used for: Black for bear hats of English Guardsmen. Cubs used for some trimmings. Used as floor throws and rugs.

Repairability: Thick, strong leather accepts repairs if not dried out; dangerous when used as a rug near heat source.

Remodelling: Very rare.

Care: See repairability.

Look Alikes: Not easy to imitate because of size. Imitations show skin joinings. Synthetic imitations are common.

Comments; History: New mounted polar bear about $2000 in 1970, depending on size. Mounting $300 plus.

Name: Beaver

Size and Appearance: Winged, oval-like skin, like most amphibians. Ranges in size two to four feet, very wide. Guard hair shiny and coarse. Fiber ranges from dark brown on back, shades off to light golden beige on sides. Western U.S. skins large, known as "blankets," sometimes up to 65 inches of length plus width, in normal raw oval state. Canadian beaver from black-brown to pale, almost silvery. Redness undesirable, in any beaver.

Geographical Sources: North America. Wild. Ranching experiments under way in Canada and U.S. Best quality from Hudson Bay area ("P.Q.") Province of Quebec.

Processing Used: Dressed: Guard hairs plucked out, leaving dense underfiber which is sheared. Used in natural and dyed brown. Dyed beige and bleached white.

Durability: 90–80% unplucked, 70–60% plucked, sheared.

Skin Layouts Used: Full or split skin. Can be let-out. Usually worked hair up, rarely horizontal.

Used For: Cloth coat trimming—jackets and coats.

Repairability: Usually good. Easier on

Beaver (natural, unplucked)

dyed than on natural sheared.

Remodelling: A let-out fur. Remodels involve stripes, etc., and therefore are costly.

Care: Has tendency to "mat." Requires at least once-a-year treatment to separate and fluff out fiber (electrifying).

Look Alikes: Pieced plates from skin cuttings. Sheared raccoon (smaller stripe) Nutria (softer leather and hair).

Comments; History: Killing of wild beaver generally under rigid legal control. Many areas restocked with beaver to balance ecology. Historically probably most important North American fur. Once used instead of money as basis of exchange. Sacs of beaver (castors) used to make perfumes. Fiber makes finest hat-felt.

Wild U.S. Catch 1967–8: 182,826. Minnesota, Wisconsin, and Alaska.

Name: Burunduki (Baronduki)

Size and Appearance: Small skin, about 3x6″ working surface. A Russian cousin of the American chipmunk. Short, coarse-haired yellow-gray background with stripes. Best sections have five dark and four light longitudinal stripes (northern Russia).

Geographical Sources: Wild. Russia, China, and Siberia mainly. Some from India have three stripes.

Processing Used: Dressed.

Durability: 30–20%.

Skin Layouts Used: Usually sewn into plate form and exported. Full skin layout, horizontal and vertical, with rounded head or "V" joining.

Used For: Use limited to novelties until late '60's and development of vogue for colorful furs—some garments, including skirts.

Repairability: Edges only, if not badly worn. Advise covering over with suitable edge binding rather than attempt to repair poor edges.

Remodelling: Not advised.

Care: Occasional wear only.

Look Alikes: Calf and kidskin imprint dyes. Also on cloth.

Comments; History: Eurasian cousin of American chipmunk, latter not used for fur at present.

Name: Calf

Size and Appearance: Young cow, usually mottled brown, black, tan, with white. Best are thin-leathered and silky moiré haired.

Geographical Sources: Northeastern Europe, Sweden originally. Now mostly U.S.

Processing Used: Dressed with extremely soft leather finish. Used in natural and dyed shades, brown, black, and beige also stencilled to imitate many other furs.

Durability: 50–40%—hair stiff will break off from friction.

Skin Layouts Used: Full skin, as cloth,

Calf (brown)

one piece to each garment section. Hair down mostly.

Used For: Sports clothes, except for black dyed.

Repairability: Edges, if slightly worn, otherwise, bind as suggested with suitable edging.

Remodelling: Not warranted because of cost-usefulness ratio. "Cut-down" only.

Care: Hair stiff, will break if subjected to abrasion. Not an everyday fur.

Look Alikes: Flat pony, dyed kidskin.

Comments; *History*: Dressing and dyeing calf skins for fur a recently developed U.S. industry. Also prominent in West Germany.

Calf (beige)

Name: Cat, Domestic

Size and Appearance: Varieties of house cat with denser fur and good leather. Varicolored spottings and some monocolors.

Geographical Sources: Netherlands, Russia, China. Holland best, followed by Danish.

Processing Used: German, Chinese domestic cats made up into plates.

Durability: 30–40%.

Skin Layouts Used: Full rectangle skin, in plates.

Used For: Sports garments, so far.

Repairability: Fair.

Remodelling: Not advised.

Care: Will shed. Wear with cloth that will not show.

Look Alikes: Bobcat, wildcat.

Comments; *History*: Russian "genet" house cat tried in U.S. some 30 years ago. Dyed to sable colors, etc. Unsuccessful. Improvement in processing that will minimize shedding is needed.

Name: Cat, Lynx

Size and Appearance: Three-quarter size of true lynx, shorter hair, less white side, with some spots on belly, darker than lynx.

Geographical Sources: U.S. and Canada —around border areas.

Processing Used: Normal yellowed flanks sometimes bleached, also dyed if poor color.

Durability: 50–40%.

Skin Layouts Used: Full and split skin. Horizontal blocked skin, also vertical.

Used For: Traditionally a trimming fur. Now being tried in garments.

Repairability: Normal edging possible.

Remodelling: Difficult because of skin size and skin pattern problems. Fur should be comparatively unworn.

Care: Same as most cats.

Look Alikes: Hungarian Cats: If dyed, will resemble American cat lynx species from warmer climates. Bobcat and wildcat smaller.

Name: Cat, Spotted

Size and Appearance: Resembles true leopards in markings. Best grades on light background.

Geographical Sources: South Asia, Africa, and South America. Best grades are the Margay cats of Argentina and Brazil. Also Texas wildcats and European cats, also known as panther cats.

Processing Used: Panther cats sheared.

Durability: 50–40%.

Skin Layouts Used: Full and split skin layouts.

Used For: Sports garments.

Repairability: Fair.

Remodelling: Only if fur and leather is fresh and change is limited.

Care: Same as other cats.

Look Alikes: Wildcat, bobcat, small leopard skins, or sections.

Comments; History: Also known as South American spotted cat.

Name: Cat, Wild

Size and Appearance: Large cat 30″ full grown. Body sandy-gray, with dark stripe down spine, and darker stripes angling vertically down sides. Ears tipped black. Sides and limbs covered with brown and black blotches.

Geographical Sources: Once abounding in mountainous and wooded areas all over U.S.

Processing Used: Dressed and sheared.

Durability: 50–40%.

Skin Layouts Used: Full and split skin.

Used For: Sports and casual garments.

Repairability: See other cats.

Remodelling: Normal for cat family.

Look Alikes: Canadian lynx—larger and lighter colored, longer-haired spotted cat, domestic cat, imitation imprints on rabbit, etc.

Comments; History: Also known as lynx cat or bobcat. Considered a predatory pest by farmers.

Wild U.S. Catch 1967–8: 10,846. Colorado and Montana leaders.

Name: Cheetah

Size and Appearance: Size of large dog. Long spotted tail. Coarse hair light brown to yellowish brown. Round black spots scattered randomly over skin. Differs from other cats in that it has long legs. Appears to have the head of a cat and the body of a dog.

Geographical Sources: Africa and Asia.

Processing Used: Normal dressing.

Durability: 40–30%.

Skin Layout Used: Full skin garments.

Used For: Fine casual and occasional garments (with trim).

Repairability: Only slight edging. Pronounced spot pattern inhibits any sizeable addition or removal of fur.

Remodelling: As with most spotted furs, very difficult. Changing nature of spot size and spacing on different parts of skin makes matching and re-shaping almost impossible.

Care: Same as other cats.

Look Alikes: Many furs bleached, dyed, and spotted to imitate, including calf, kidskin, rabbit and mink.

Comments; History: Live cheetah tamed and used for hunting. Said to be the fastest animal.

Name: Chinchilla

Size and Appearance: Larger, round rat, fur one-half to three-quarters inch long. Fur cigar-ash mottled blue-gray, changing to white underneath. Best skins blueish.

Geographical Sources: Originally wild —Chile, south Peru, and Bolivia. Now ranch raised in U.S.

Processing Used: Natural dressing— some "brightener" (blueing) usually added to enhance blueish tone. Some high fashion houses dye it green and other shades.

Durability: 30–20%. Fur extremely soft and warm.

Skin Layouts Used: Originally worked let-out in 1940's. Now worked in block form, head to head, rump to rump, into rows, both horizontally and vertically.

Used For: Trimmings on other furs and cloth. Small shrugs and shoulder capes, rarely on larger capes, jackets and coats.

Repairability: Edges repairable if not too worn. Leather very thin, difficult to handle.

Remodelling: See above. Warranted if condition of fur and leather is good enough. Fine, weak remodelled garment should be hand stayed.

Care: Obviously for occasional wear. Do not allow to become matted. Blow out in front of fan, etc., or any air-blowing device.

Look Alikes: Chinchillone—longer more yellow. Chinchilla (American) rat "Bastard." Chinchilla sub-species from

Chinchilla

lower altitudes. Also rabbits dyed to simulate chinchilla. Natural gray rabbits raised to imitate.

Comments; History: Exported 1829 to 1914 until nearly extinct. Three governments banned trapping and export in 1914. M. L. Chapman, American mining engineer, smuggled eleven live specimens out in 1924. He started chinchilla ranching in U.S. First assortment marketed in 1949. Family finally sold out farm in 1970.

Name: Civet Cat

Size and Appearance: Approximately half the size of skunk. Broken black and white lines in lyre formation. Zoologically a true skunk with same defenses.

Geographical Sources: Most of U.S. "Bucktail" north central U.S. Narrow

white stripe, white tail. Western and southern U.S. broader stripe.

Processing Used: Used natural if good color and lyre pattern, otherwise dyed to imitate other furs.

Durability: 50–40%.

Skin Layouts Used: Mostly skin-on-skin with "V" and rounded-head joinings. If dyed can be worked, block horizontal and vertical.

Used For: Dyed form traditionally used for trimmings. Returning to popularity, with natural design emphasized for garments.

Repairability: Fair, for edging—good if dyed monotone.

Remodelling: Prominent pattern presents usual problem common to patterned furs, cats, etc.

Care: Normal

Look Alikes: Chinese Civet. Yangtse Valley China, "Bush Cat." Fur fairly wavy and silky. Imprint dyed rabbit.

Wild U.S. Catch 1967–8: (combined with skunk) 19,459. Michigan and Virginia leaders.

Name: Dog

Size and Appearance: Dark brown—black color. Occasionally spotted. When live looks like large chow dog. Long silky hair most desirable.

Geographical Sources: Wild and domestic. Mainly from China, usually Manchuria. Used in Korea as a beast of burden.

Processing Used: Dressed. Bleached and dyed to many shades.

Durability: 70–60%.

Skin Layouts Used: Full and split skin.

Used For: Trimmings, when dyed, also naturals; Parkas, throws, etc.

Repairability: Normal.

Remodelling: Not advised.

Care: Normal.

Look Alikes: Other long-haired dyed furs, "Moufflon" (sheep), etc.

Comments; History: Skins bleached and dyed commercially in 1920. Dye came off on workers hands, causing objections to handling. Fur is rarely used today due to its bulkiness and excessive weight.

Name: Ermine

Size and Appearance: Little, long slender animal. Natural color in winter, white and black-tipped tail. In summer turns tawny brown—"summer ermine." Working length about 15–16".

Geographical Sources: Siberia (best). Canada, Alaska, north Russia, some from China.

Processing Used: Dressed. Seldom pure white, especially near urinary organs. Requires bleaching and rebleaching. Summer types tipped or dyed to deepen color.

Durability: 60–50%.

Skin Layouts Used: Coats of 1920's once worked semi let-out. Now handled

Ermine

split skin in horizontal, vertical and chevron layouts as well as in let-out form.

Used For: In white, traditionally a most formal fur. "Summer ermine" used for more casual garments. White is still the more popular color.

Repairability: Fair. Avoid attempting to replace for matching; material not easily obtainable.

Remodelling: Cost of fur warrants re-use if leather and fur are in good condition. Should be reinforced. Note: reblending or rebleaching will weaken leather and stitching.

Care: Keep white ermine out of sunlight, preferably in dark place when not using.

Look Alikes: Some "summer ermine" actually U.S. weasels. Also sheared and grooved white rabbit. Also Manchurian ermine—yellowish, orange tone.

Comments; History: Name derived from Armenia, shipping center in time of Homer.

Name: Fisher

Size and Appearance: A large marten, resembles a large black cat, blackish, brown, gray at head and neck. American cousin of "Russian Sable." Differs from other martens, has no throat patch. Females more valuable though smaller, have soft silky fur. Length up to 30″ (Male), tapered width.

Geographical Sources: Larger lighter-colored skins in western Canada and U.S. Smaller, darker, silkier skins from eastern Canada.

Processing Used: Normal. Poor skins blended or even dyed sable shades.

Durability: 100–90%.

Skin Layouts Used: Split skin as trim. Let-out for garments.

Used For: Traditional use was for now unfashionable scarfs. Now used for collar trim and full garments.

Repairability: Some re-edging possible.

Remodelling: Throat coloration, plus neck and shoulder formation, limits remodelling to leaving this area unchanged. Pattern must be selected with care.

Care: Normal. Strong fur.

Look Alikes: Natural; martens, dyed, sables, foxes, bassarisk.

Comments; History: Size not a factor in value. Smaller pelts are silkier, better in color.

Wild U.S. Catch 1967–8: 1,139. Only two states; New Hampshire, New York.

Name: Fitch

Size and Appearance: Medium-sized, slender tapering body. Underfiber yellowish, top hair ranging from white to brown. Best have long silky evenly colored black top hair and dense underfur. Black marking on lower body is distinctive.

Geographical Sources: Two main groups: White (Siberian) dark (Germany, Austria, Poland—best), Russian whitest. Bright yellow "Paradise" fitch from Mongolia.

Processing Used: Dressed: Used natural now. Once dyed in mink and sable shades when less expensive.

Durability: 70–60%.

Skin Layout Used: Originally vertical in natural or dyed, skin-on-skin or some let-out. Natural now worked exclusively in "V" chevron and straight horizontal, both split skins.

Used For: Formerly popular in natural form as trimmings, also in dyed form. Now used almost exclusively in coats and some jackets, all in natural form.

Repairability: Edging possible. Sharp color variants prevent "patching."

Remodelling: Warranted when fashionable, but color and horizontal layouts limit pattern changes to length of skin. In 1970 desired lengthenings difficult for this reason.

Care: Normal. Must be cleaned regularly to conserve whiteness. Five-year-old plus garment may need slight bleaching.

Look Alikes: Tiger-fitch (Perwitsky) from Poland has tiger-like stripes. Raccoon and muskrat in semi-bleached form resemble fitch.

Comments; History: Whitest, most contrasting skins most desirable. Breeding attempted in Vancouver and Washington in 1940's. Cousin to the American skunk, same scent. Fluctuates widely in value according to relationship of fashion to supply. In 1970 white Russian fitch unobtainable and much in demand, shot price above that of most mink.

Fitch

FOX FAMILY

This family of animals is found all over the world, both wild and ranched. Some varieties remain fashionable from generation to generation; most, like silver fox periodically rise and fall from fashion's favor. The following represents those which are being used or were used for fur in the last half-century.

Fox ranching, like mink ranching, has progressed to the point where it has produced and developed mutations. In the past it was profitable to improve the appearance of some types of foxes by judicially gluing in contrasting colored badger hairs in a seemingly natural pattern to simulate expensive silver fox. This technique, called "pointing," is now largely obsolete because of its costliness and the drop in price of the originals.

U.S. Wild Catch 1967–8: 138,629 for all species; Iowa and Wisconsin leaders.

Name: Blue Fox

Size and Appearance: Color phase of white fox. Blue fox is darkish blue-brown in winter, more brownish than in summer. Best grades are palest with most blue. Shadow blue, white fox with blue shade on spine developed in 1950's.

Geographical Sources: Introduced to Pribilof Islands by Russians and spread over the Islands. Original wild sources include Greenland and northern Europe. Ranched expensively in Norway, some in Canada and U.S. In 1970 one-third of world production from Scandinavia.

Processing Used: Normal. Some off-color skins bleached slightly to enhance lightness.

Durability: New 60–50%. Older skins develop brittle and weak leather.

Skin Layouts Used: As a trim, usually full or split skin. In garments, either horizontal, split skin or hair down let-out.

Used For: Cloth and fur coat trimmings. Also for small one- and two-skin shoulder shrugs. Also jackets, coats, floor length capes.

Repairability: Fair. Edges wear if fur is matted, as with most foxes.

Remodelling: Slight changes only. Coloration and size, also possible leathering inserts limit changes. Never on old pieces. See durability.

Care: Keep fur fluffy by blowing and regular cleaning.

Look Alikes: Dyed flying squirrel. Moufflon, sheep, and long-haired rabbit dyed to imitate it, also cheaper foxes.

Comments; History: In 1925 blue fox

Blue fox

raising in Alaska important enough to warrant U.S. government bulletin. Most now come from Norway. Several governments killed off whites and protected blues, allowing latter to become dominant. First sale of Shadow Blue in 1960's brought $400 per skin.

Name: Fox, Cross

Size and Appearance: Originally a natural color phase ("Wasserman sport") breed of wild red fox. Basic color: red tinted with yellow—some show tinges of silver. Outstanding characteristic is pronounced darker-haired "cross" formed by dark hair across shoulders and again just below eyes and ears. Fur above and below crosses lighter than rest of pelt.

Geographical Sources: North America,
and all countries with red foxes.

Processing Used: Natural.

Durability: 60–50%.

Skin Layouts Used: Full skin and split skin.

Used For: Trimmings, small shrugs, jackets and occasional longer garments.

Repairability: Fair.

Remodelling: Coloration a drawback.

Care: Same as other furs.

Look Alikes: Some degree of "cross" configuration found on all red foxes. "Tanuki" from Japan—many other names —dyed, resembles dyed fox.

Comments; History: Another fur which once had high fashion appeal but is now unfashionable.

Name: Fox, Gray

Size and Appearance: Grown fox basically silver-gray, slightly tinged with red. Skin gives mottled appearance due to gray and black hair. Flanks off-white, with reddish shading.

Geographical Sources: Mostly American. Best grades from northern states.

Processing Used: Natural. Some tipped slightly to add another "color tone." Black, red, etc., poor colors dyed.

Durability: 60–50%.

Skin Layouts Used: Full and split skin. Block type.

Used For: Trimmings on popular priced cloth coats, also for casual garments, usually in horizontal block form.

Repairability: Normal for foxes.

Remodelling: Not advised, does not warrant cost.

Care: Normal fox care.

Look Alikes: Other foxes if dyed.

Name: Kitt Fox (Corsac Fox)

Size and Appearance: A full-grown small fox. All have short, soft hair and little top (guard) hair. All belly fur yellow. Guard hair yellow with white tips.

Geographical Sources: Best from Russia and Siberia (Corsac Fox). North American type "Swift" foxes, best are Canadian. Poor grades from Mexico and American southwest "Desert Foxes" also South American gray fox misnamed Kitt Fox.

Processing Used: Dressed. So far used in natural shades, or can be dyed.

Durability: 50–40%.

Skin Layouts Used: Mostly block, horizontal, for jackets and coats. Some cloth coat trim and linings.

Used For: Trimmings, popular-priced cloth coats. Also "young" fur garments.

Repairability: Normal for foxes.

Remodelling: Probably not warranted.

Care: Normally fox care.

Look Alikes: Desert foxes from southwestern Asia, India, Afghanistan, Iran, and Arabia.

Name: Fox, Platina

Size and Appearance: Strain of silver fox, with lighter platinum-gray color and full white head. Some strains similar to white neck silver.

Geographical Sources: Norway originally. Breeders exported to other countries.

Processing Used: Natural. Whiteness may be slightly enhanced by slight bleaching.

Durability: 60–50% new—20–30% old.

Skin Layouts Used: Same as other foxes, silvers or blues.

Used For: Originally used for mounted scarfs when these were popular. Occasionally for fine trimmings, shrugs, and garments.

Repairability: Fair.

Remodelling: See other foxes same group.

Care: Similar to other foxes.

Look Alikes: "White Neck" silvers.

Comments; History: First shipment from Norway in 1939. One skin sold for $10,000. First garments made by Molyneux for Duchess of Windsor and Princess Lucinge. Faucighy mutation of silver fox, discovered and developed in Norway. Development through ranching hindered by loss of fashion acceptance. Once again finding favor in the new fashions.

Name: Fox, Red

Size and Appearance: Comparatively bright-red coloring. Long bushy white-tipped tail. Red fox like all its color phases, has characteristic "cross" with lighter areas above and below shoulder "cross." Long thin head, becomes wider toward rump. Best grades 35″ from nose to tail base, complete flow of top hair, cushion full under fiber.

Geographical Sources: All continents except South America. Clearest reds from U.S. and Canada. Kamchatka (Siberian), red darkest (cherry red).

Processing Used: Dressed: Poorer color, good hair skins dyed black, amber, gray, etc., to imitate currently popular fur shades.

Durability: 60–50%.

Skin Layouts Used: Once popular for mounted scarfs. Now used for hats, jackets and coats, usually split skin horizontal. Can be let-out, but not through "cross." Fur popular for casual garments as trim, jackets and coats, especially if dyed color is popular.

Repairability: Fair. Edges can be reworked. Some patching possible.

Remodelling: Possible with conservative changes. Not too practical for radical styling, either block or let-out.

Red fox

Care: Will not stand much friction. Keep fur fluffy.

Look Alikes: Mongolian, light, dull hair, poor reddish color.

Comments; History: Ecological imbalance has increased numbers in 1960's throughout North central states, also Manitoba, Saskatchewan. Fur is now in extreme demand. Supply is limited once again because of viral disease which periodically limits supply.

Name: Fox, Silver

Size and Appearance: Mutation of red fox family, blue-black fur sprinkled with silver strands. Tip of tail is white. Guard hair white, band width varies. Best grades have full silver and clear black grotzen line.

Geographical Sources: Entirely ranched. Canada, U.S., Alaska, Russia, and Norway.

have all tried ranching at one time or another.

Processing Used: Dressed: Used natural. Skins with insufficient silver once "pointed." Silvery badger hairs glued in to give better appearance.

Durability: 60–50% new. 20–30% old.

Skin Layouts Used: 1940 Jackets featured re-set technique (twin skins from one). Now used in full or split skin. Leathered.

Used For: Trimmings, edging, hats, shrugs, some garments.

Repairability: Fair when new.

Remodelling: Full silver not too difficult to rework if leather is in good condition. Depends on price structure.

Care: Clean yearly and keep fluffy.

Comments; History: Ranched developments include platinum, ring-neck platinum, pale platinum, white marked silver. Largest U.S. rancher was Fromm Bros. who had 36,000 live foxes on Wisconsin farm in the late '30's. Another casualty of extreme fashion changes. In 1970 practically out of production due to lack of market. Old jackets retrieved from attics, antique fur shops found favor with young in 1967–9. Now finding renewed favor in trimmings, hats and shrugs. Rarely used in garments.

**Silver Fox Tuxedo
(coat is broadtail processed lamb)**

Name: Fox, White

Size and Appearance: Very heavy, dense under fiber. Top hair not easily seen, same shade as fiber. Side skin has full top hair. Variation shadow blue fox developed in 1950's. White with blue shade along spine.

Processing Used: Involves prevention or counteracting yellowish tinges, including bleaching if necessary.

Durability: 50–40%.

Used For: Used in past for more formal evening garments, shrugs.

Repairability: Good.

Remodelling: Good—monotone color lends itself to manipulation.

Care: Susceptible to rubbing wear. Clean and fluff out.

Look Alikes: Dyed flying squirrel—white hare.

Comments; History: Wide fluctuations in popularity and demand. Now in favor. Wild stock killed off by some governments to encourage growth of blue fox stock in its place. "Paradise Fox" discontinued name for white fox with artificially glued (pointed) black hair. Must now be labelled "Pointed."

Name: Guanaco—Guanaquito

Size and Appearance: Variety of small camel about size of pony, but has no humps. Pelt most popular in young, small state, Guanaquito. Red, brown long-haired back with sharply defined whitish flanks, belly and under legs, showing sharp pattern

Guanaquito a young guanaco two-three months old.

Geographical Sources: Nearly extinct in most of lower South America, Argentina has some wild on its pampas.

Processing Used: Normal. Very often dyed if color poor. May be plucked, or plucked and sheared occasionally.

Durability: 50–40%.

Skin Layouts Used: Skin-on-skin, rounded or "V" head-to-rump joining to emphasize pattern.

Used For: Trimming, slippers, boot trim, sports jackets and coats. Rarely used as fur today.

Repairability: Fair. Can be re-edged.

Remodelling: Limited by sharp pattern.

Care: Requires brushing of hair to counteract matting. Tends to curl when wet, needs ironing by specialist—not by layman.

Look Alikes: Dyed fox, dyed lamb. Llama—Can carry load of 75-100 lbs. Camel, black and spotted in color. Alpaca—Bigger head, shorter thicker neck, long heavy fleece. Vicuna—Rare, small, soft. Woven into cloth called "compi" as lustrous as silk.

Comments; History: Small guanaco misnamed "vicuna" an entirely different animal whose fiber makes one of the world's fine cloths.

Name: Hamster

Size and Appearance: One of smaller fur-bearing animals. Maximum length 12", many less. Many colors on flanks and longer body, white and yellow on sides. Color varies considerably, some all black.

Geographical Sources: From Hartz Mountains and sandy areas from north Germany to Siberia. U.S.S.R. and Germany.

Processing Used: Dressed and sewn into plates.

Durability: 40–30%.

Skin Layouts Used: Home-sewn into plates, 60 to 75 skins, in rectangles.

Used For: Sports clothes, dresses, suits, etc.

Repairability: Poor. Fur is not strong. Matching must be by full skins, which becomes expensive.

Remodelling: Not advised. Too difficult and costly for return in wear.

Care: Not for daily wear. Avoid friction.

Look Alikes: Rabbit stencilled imitations.

Comments; History: Pest in some parts of Germany, steals grain. Rarely used as fur in this country.

Name: Hare

Size and Appearance: Larger than rabbit, light in color and with much longer hair.

Geographical Sources: Siberia and Scandinavia. Imported live to Siberia.

Processing Used: Dressed and often dyed.

Durability: 20–10%. Also sheds.

Used For: Once used occasionally for children's garments. Now discontinued due to increased use of rabbit. Inexpensive shrugs. Toys.

Repairability: Fair.

Remodelling: Not advised.

Care: Sheds. Use vacuum cleaner. Wear with complimentary colors.

Look Alikes: Long-haired ranched Belgian, French rabbits, larger than other rabbits. Dyed and sheared to black and other shades.

Comments; History: Used as fur. All but discontinued at present in U.S. for reasons above.

Name: Jaguar

Size and Appearance: Larger cousin of Leopard. Markings differ in that rosettes are pointed, not round, and much larger. Small spots inside each marking. Center back has straight line marking. Marking in rows parallel to center line. Spots blackish, background color varies, lighter to darker shades of tan.

Geographical Sources: Once ran wild from southwest United States through much of South America.

Processing Used: Dressed, with emphasis on thinnest possible leather for garment use. Otherwise, for trim, rugs and wall hanging.

Durability: 60–50%.

Skin Layouts Used: Almost entirely full skin hair down in garments. Split skin for trimmings, also very rarely as novelty horizontal effect.

Used For: Novelties, fine trimmings, garments, jackets, coats, rugs.

Repairability: One of those furs very difficult and expensive to patch because of pattern. Bad edges may show if cut down. Consider binding edges.

Remodelling: Only for cut-down version. Sections of fur garment cannot be remodelled because of pattern. Warranted where possible because of value, but measurements of new pattern must be checked against fur skin-by-skin.

Care: Hair not soft. Has tendency to snap off under friction. Not for daily wear.

Look Alikes: Ocelot and some leopards. Prints on calf, kidskin, lamb, rabbit and mink, latter especially prominent in 1969–70 season.

Jaguar (sable trim)

Comments; History: Not now permitted in this country for conservation reasons, although used in many foreign countries.

Name: Kangaroo

Size and Appearance: Working length of grown skin to six feet. Two major species in gray and red tones, lighter to whitish sides.

Geographical Sources: Australian—red and gray species.

Processing Used: Normal—some poor-colored skins dyed.

Durability: 40–30%.

Skin Layouts Used: Full skin.

Used For: Sports and casual garments.

Repairability: Requires frequent re-edging. Sheds.

Remodelling: Not advised.

Care: Maximum.

Look Alikes: Wallaby—very small edition of Kangaroo.

Comments; History: Regarded as pest in parts of Australia as late as 1973.

Name: Kidskin

Size and Appearance: Young goat. Some have markings closely resembling broadtail. Older kidskin has wavy hair. Very young or unborn termed "Golyaks." Chinese skins gray with black, thin grotzen African skins mottled, multi-colored. Gray, tan, black and white combinations.

Geographical Sources: Mainly Chinese. Also India and Africa. Chinese best.

Processing Used: Dressed: Chinese gray skins dressed natural. Others dyed brown and light-brown shades, occasionally black.

Durability: 30–20%.

Skin Layouts Used: Traditional on gray; round-head joining on split vertical or horizontal joinings. Also worked with concealed seam. Made into plates in China. Often sewn by hand.

Used For: Jackets and coats.

Repairability: Required often. Edges can be cut down, if not dye-printed in some exotic pattern.

Remodelling: Questionable in view of cost against potential use. Multi-colored types difficult to handle in remodel.

Care: Not for daily wear. Avoid friction.

Look Alikes: Dyed calf.

Comments; History: Nearly full-grown skins with top hair removed once sold as "Moufflon."

Name: Kolinsky

Size and Appearance: Originally from Kola Peninsula in northern Russia. Smaller, shorter-haired weasel cousin of mink. Naturally yellowish-brown coloring.

Geographical Sources: True Kolinsky from Siberia, Russia, or Manchuria.

Processing Used: Almost invariably dyed to resemble mink or sable.

Durability: 50–40%.

Skin Layouts Used: Rounded or "V" head-rump joining, split skin chevron. Best grades may be let-out. Also split skin horizontal and vertical.

Used For: As replacement or substitute for mink where possible. Trimmings, scarfs and jackets.

Repairability: Fair. Will take re-edging.

Remodelling: Difficult to secure added material if needed.

Care: Normal. Needs to be kept fluffy.

Look Alikes: Chinese weasel (China mink) hair shorter.

Comments; History: Usually sold without tails, hairs of which are used for artists' paint brushes.

171

LAMB FAMILY, INTRODUCTION

Many varieties of lamb from all over the world now used as fur. The most popular is Persian lamb (karakul) originally perfected in Bukhara, Russia, more than 200 years ago. Now raised extensively in Rumania, Afghanistan, Southwest Africa and South Africa, Russia. Now being raised in Persia for the first time.

This fur also classified by the age of skin at time of taking.

"Broadtail" stillborn, or unborn lamb if mother has died. Very soft thin leather and very fine short hair in moiré pattern.

"Persian broadtail"—pelt one or two days old. Leather slightly thicker. Best have "pine" tree pattern with slightest suggestion of curl.

Persian lamb. Up to week old. Hair has grown to full curl. Leather thicker. South African skins have flatter curl, thinner leather, more moiré pattern than the more regularly curled European and Asiatic karakuls. South African skins have shorter necks than the European lamb.

Other lamb varieties used as fur include:

Caracul—lamb with wavy or incomplete curl. Mostly from Central Asia and China. Best known assortment "Chekiang lamb."

"American broadtail" dyed lamb. Young Lincoln lamb from Argentina which shows natural moiré pattern when sheared. Dyed to simulate the Russian broadtail.

"Mouton lamb"—dyed sheared sheep.

"Mongolian lamb"—long wavy shiny hair 3–4 inches long usually sold in natural or bleached white.

Sueded lamb—any lamb whose leather is finished off and glossy so that it can be used as outer surface of a garment. "Shearling"—sheared lamb, is often used this way. Very popular the last several years.

Kalgan lamb—small, tightly-curled white lamb from Asia.

Yemen lamb—brown and white wavy small lamb. Also "Borregos," African, Austrian and other varieties now being developed and marketed. Natural or sheared forms of various lambs.

Tibetan lamb—shorter-haired variant of Mongolian lamb.

Name: Caracul

Size and Appearance: Means "Black lake" in Russian. Also name of town in Russian Turkey. Small squarish skin, nattural black, brown or white. Has wavy, shiny hair. Leather thin and soft.

Geographical Sources: Lower central Asia, Mongolia, and Manchuria. Chekiang province best grade and best known.

Processing Used: Dressed and usually dyed in dark colors, brown or black or gray. Dyeing to light colors weakens fur.

Durability: 50–40%.

Skin Layouts Used: Concealed scallop or wavy joining to give uniform appearance. Sometimes made into plates in country of origin.

Broadtail Processed Lamb

Russian Broadtail (sable collar)

Indian Lamb

Persian Lamb (lynx collar)

Used For: Garments. Very popular between World War I and II.

Repairability: Excellent. Edges can be cut down without showing. Additional material easy to match in, if available.

Remodelling: Limited by condition of fur leather. Can be remanipulated very well.

Care: Dressy, not to be subjected to hard usage.

Comments; History: Cuttings from skins, heads, paws, etc., also made into piece plates. Sometimes confused with "kid-caracul," young goat of India, China, Africa. Name often confused with "Karakul."—See African persian lamb.

Name: Mouton Lamb

Size and Appearance: Large sheared sheep. Natural color usually off-white. Skins very large, sold by square feet 5′x4′, 6′x4′, etc. Quality in direct relationship to lightness and softness of skin.

Geographical Sources: Most good strains of dense fiber sheep, especially merinos from U.S., Australia, South America, etc.

Processing Used: Dressed and dyed to brown, black and other shades. Sheared and treated with synthetic plasticizers to prevent curl and enhance luster. Also stencilled to imitate other furs. Dyed to simulate Alaska seal and beaver.

Durability: 70–60% (most shorn lambs).

Skin Layouts Used: Worked as cloth. Skins large enough to be cut to pattern lines. Occasionally smaller skins split and worked horizontally, sometimes with leather inserted.

Used For: Mostly coats and jackets—some novelties. Also for self-lined garments. A best buy for utility and warmth without being expensive. Also used for toys, slippers, linings, etc.

Repairability: Poor—re-edges and repairs will show seams.

Remodelling: Not difficult, but cost may not warrant expense of remodelling unless fur is in excellent condition and is made into a small garment.

Care: Normal.

Look Alikes: Pile fabric.

Comments; History: Recent experiments with sueding leather of furs of lamb family to commercialize age-old Hungarian shepherd's coat, indicate possibility of making reversible self-lined sheep coat. In U.S. shearlings (natural sheepskin with wool left on) were shown in reversible, washable, ski and aprés-ski garments in 1970, an event so important as to elicit the approval of the U.S. Department of Agriculture.[3] Very popular to this day.

Name: Lamb, Persian (Karakul)

Size and Appearance: Black, fat-tail lamb, ten days old or less. (See Introduction: Lamb Family) with tight, close full curl. Naturally black, gray, brown, and white. New colors being bred in U.S.S.R. include "Sur"—dark base and light tips, "Goligaz"—marbled effect in gray or pink; "Halili"—black with brown saddle; "Sedendi"—frosted with white hairs.

Geographical Sources: U.S.S.R., Afghanistan, Southwest Africa, South Africa. Southwest Africa produces 70% of African karakul. In 1969, all of South Africa produced 5.3 million pelts.

Processing Used: Skins pre-salted raw, then dressed. Most black, dyed black. Now being dyed in blue, red, and various "tip" shades. Colors involving pre-bleach may weaken fur and leather.

Durability: 60–50%.

Skin Layouts Used: Almost entirely one form or other of the concealed seam, occasionally interspersed with ribbon or leather. Gray persian worked let-out in concealed seam.

Used For: Most fur garment uses, trimmings, novelties.

Repairability: Best of all furs.

Remodelling: Best of all furs. Additional material easily added. Areas can be moved without difficulty. Only when furs start to peel is fur too old for remodelling.*

Look Alikes: Other curled lambs. Also pieced plates from same fur including paws, heads, pieces, all made into plates.

Comments; History: Some confusion as to origin. Most consider Iranian plateau (of Persian Empire) source. In 1949 F.T.C. ruled that "persian lamb" can come from other countries.

* Leather consists of three layers: curl layer, sponge layer and back layer. When old, top layer separates (peels) from spongy layer, because of its drying the leather peels or cracks. This can sometimes be partially corrected by re-oiling the leather. This can, however, be prohibitively costly because it involves the complete remaking of the garment which will still not be new.

Name: Leopard

Size and Appearance: Somali skins are pale, silky, flat 30″–40″ working length. Spots dark in contrast to rosy tan background. Spots small, grouped in rosettes. Chinese skins larger heavier, larger spots, reddish-brown tinge. Ceylon (India) skins flat and silky but brownish.

Geographical Sources: Asia and Africa. Best skins with shortest hair; best coloring, smallest spots from areas near Equator. Somaliland best. Ethiopia, Kenya, Tanganyika also desirable. India, Ceylon, China and Central Africa usually heavier, tawny red in color. "Snow" leopard from Asiatic mountain ranges of Himalayas; long-haired grayish-white skin, with clear white flanks and much longer hair than other leopards. "Clouded" leopard from Formosa, mixed angular large markings, grayish tone.

Dressing: Normal.

Durability: 80–70%, if not subject to excessive friction. Hair is brittle and has tendency to break.

Leopard

Skin Layouts Used: Full skin hair down for most long garments. Split skin for trimming or occasional split skin layout.

Used For: Coats, trimmings, jackets and novelties.

Repairability: Slight edging repair feasible, although wide cut may show. Patching almost impossible as for most patterned furs.

Remodelling: Limits as indicated for large cats. Remodelling warranted at almost any cost if leather and fur in good condition. Skin pattern layout must be aligned to remodel pattern lines.

Care: Should be worn with care to prevent friction.

Look Alikes: American leopard cat, Ocelot (smaller with elongated spots). Leopard prints on calf, kidskin, mink and rabbit, also Asiatic leopard catskin plates.

Comments; History: Fur is prohibited from import into this country for conservation reasons. Still used in many foreign countries.

Name: Lion, Mountain (Cougar)

Size and Appearance: Tan or tawny in color, darker on the back, lighter on the under parts. Hair is short and soft. Largest Arizona catch ever recorded—276 lbs. Also known as "Catamount," "Panther," "Puma."

Geographical Sources: Mostly Arizona, some in neighboring states—few in Florida.

Processing Used: Dressing: Normal.

Durability: Unknown. Short soft hair probably not durable.

Skin Layouts Used: Large skin, cut to pattern like cloth.

Used For: Almost never used for garments.

Repairability: Unknown.

Remodelling: Unknown.

Care: Normal.

Look Alikes: Light calf, tawny shades sheared pony.

Comments; History: Can and does kill full-grown horses. Arizona pays bounty. Came into use for fur in 1967–1968 season. Rarely used as fur.

Name: Lynx

Size and Appearance: Long, thin, like cat family. Longish light-colored body, spotted slightly. Silky hair. Fur dense and soft in varied shades of light brown. Best grades must have clear color bellies without brownish tinge. Good skin above 36" from nose to tail base. Tail tipped with black hairs above and below. Feet large.

Geographical Sources: Russia—almost white, light spots. Smaller, coarser, lacks white flanks. Canada—considered best—spots blueish in tone. Scandinavia, northern Asia (to Himalayas), U.S. (Alaska).

Processing Used: Dressed normally. Sides bleached if off color.

Lynx

176

Durability: 60–50%.

Skin Layouts Used: Split skin, full skin. Can be let-out. Layouts that will feature sides rather than grotzen, part of which ("mane") is cut away.

Used For: Major use is for trimmings on cloth and fur garments.

Repairability: Fair if no material is needed.

Remodelling: Can be reworked as a trimming, within limits. Otherwise strong coloration pattern is drawback.

Care: As cat, requires care to prevent loosening hair. Clean regularly.

Look Alikes: Lynx Cat: Smaller, tail tipped black on top only. Also dyed foxes to simulate; also long-haired printed rabbit, sheep, etc. Pale wolves.

Comments; History: Conservation depends on ample supply of rabbit meat for reproduction.

Wild U.S. Catch 1967–8: 1,688. Almost all from Alaska.

Name: Marmot

Size and Appearance: Large rodent. Blueish in tone before hibernation (about November). Yellowish post hibernation skins show hair swirl due to turning in sleep. Hair coarse and sparse. Unique because hair flow of head area is up toward snout.

Geographical Sources: Siberia, Mongolia, Manchuria and Russia. Mostly Chinese.

Processing Used: Almost always dressed, dyed and striped to mink tones. Best occasionally let-out. Blues used natural in 1969 fur coats.

Durability: 60–50%.

Skin Layouts Used: Most used skin-on-skin, with zig-zag joinings and striped. Fine skins used for let-out to imitate mink.

Used For: Jackets, coats and popular priced cloth coat trimmings.

Repairability: Fair, if in dyed and striped form, problems similar to real let-out garments.

Remodelling: Leather has tendency to soften as it ages. In general radical, expensive, remodelling not warranted.

Care: Subject to hair breakage due to friction.

Look Alikes: Tarbagam marmot of Mongolia, heavy, large, full, long hair. Susliki marmot—"Peschanik" short, thin hair, tawny color, also mink-dyed striped rabbit. Also viscacha.

Name: Marten, American

Size and Appearance: Size of small mink. Fine dense underfur. Guard hairs as long as a fox. Varies in color from blue-brown to chocolate brown, from pale brown to yellow with orange tones. Extra large skins 22–25" long.

Geographical Sources: American cousin Russian sable, Alaska, Canada and U.S. Best grades from Yukon, Mackenzie River, North Hudson Bay, Labrador.

Processing Used: Most skins blended or tipped. Yellowish, lighter skins tip-dyed to sable shades.

Durability: 60–50%.

Skin Layouts Used: Full skin. Also split skin for trimmings. Can be let-out if needed.

Used For: Scarves and Boas, jackets and some coats.

Repairability: Fair. In garment, edging possible if not too deep.

Remodelling: Usually in let-out form,

in which case has usual drawbacks. Remodel costs warranted because of value of fur.

Care: Needs to be kept unmatted and fluffy, especially in boa and scarf form.

Look Alikes: Kojah long-haired mink,

stone and baum martens if dyed.

Comments; History: Known as "Hudson Bay sable" before F.T.C. regulations.

Wild U.S. Catch 1967–8: 8,845. Mostly from Alaska.

Name: Marten, Baum

Size and Appearance: Small animal like American marten. Has lighter sometimes yellowish throat markings. Naturally gray, brown. Fine tail most desirable. Not as fine or silky as stone marten or sable. Some yellowish.

Geographical Sources: Northern Europe, Asia and Asia Minor, Himalayas.

Processing Used: Normal. Occasionally blended or dyed if poor color. When dyed to sable shade, almost undistinguishable from sable.

Durability: 60–50%.

Skin Layouts Used: Full skin for scarfs, split skin for collars let-out for longer garments. Europeans and veterans use "V" type let-out. Others use mink split skin form same as American marten, fisher and sable.

Repairability: Worn sections can be cut away on large garments needing re-edging. Claws, tails, etc., replaceable on scarves.

Remodelling: Same as other martens.

Care: Same as other martens.

Look Alikes: Jap or canary marten. Smaller, shorter fur. Other martens if in dyed form.

Comments; History: Once known as pine marten, American varieties still known by that name in some states.

Name: Marten, Stone

Size and Appearance: Not as fine or silky as sable or baum marten, very fine tail, distinct because of white patched neck which forms "V" when seen from underside.

Geographical Sources: Area around Caucasus Mountains, Asia Minor.

Processing Used: Normal. Some blending if off color. Also dyed sable shade.

Durability: 60–50%.

Skin Layouts Used: Full and split skin can be let-out.

Used For: Originally most used for scarves. Now less popular, since this use unfashionable. Some trimmings and jackets.

Repairability: Edges repairable on gar-

Stone Marten

178

ments. In scarf form—feet, tail can be repaired or replaced.

Remodelling: White patching at throat —limits flexibility of change through this area. Otherwise presents same problems as do all let-out garments. Warranted if fur and leather is in good condition.

Care: Avoid matting. Clean when any sign of flat matted hair areas show.

Look Alikes: Other martens, especially if dyed and throat patch removed.

Comments; History: Tails very desirable for decorative purposes on other furs and as decorative trim.

Name: Mink

Size and Appearance: Large weasel. Original wild color rich dark brown. Now extensively ranched, resulting in range of color from white to dark brown. Long narrow body to 26″ long. Guard hairs ¾″ to 1″, should be 5/16″ longer than underfiber on female, ½″ longer on male.

Geographical Sources: Originally North American. Now ranched in many countries, usually from North American breeding stock. Best wild sections: Labrador, Nova Scotia, Quebec, New England.

Processing Used: Dressing natural or "color added." See F.T.C. regulations. Some colors partially bleached to enhance color lightness. Skins with good hair and fiber structure but poor color dyed in various colors, black to bright pastel colors. Also bleached and sometimes sheared and dyed.

Durability: 90–80% ranched. 80–70% wild.

Skin Layouts Used: Skin-on-skin, split skin, and let-out. Pieces assembled in many intricate patterns.

Used For: Mink used for every possible type of fur garment, plus novelties, earrings, clips, suits, trimmings, etc. All parts of skin used.

Repairability: Excellent. Edges and worn spots can be repaired. Pieced minks sometimes more difficult, especially if boldly patterned.

Adaptability of mink. Three ways to show White Mink (top to bottom): cape, full-length coat, blouse-jacket.

Remodelling: Mink garments can be remodelled to any reasonable change, if fur and leather is in good condition. Latter often dries up and disintegrates when wet if too old. Fur can be added, color restored by blending. Good remodel is expensive.

Care: Requires only normal care. Should not be subject to friction of metal bracelets, etc. Regular cleaning will prolong life and looks.

Look Alikes: Many weasels, Japan and China mink, dyed marmot, rabbit, muskrat.

Comments; History: Small yet dangerous when cornered, long jaw and sharp teeth require careful handling when ranching. Most new ranched shades bring very high prices when first introduced, then find level. New shades and hues developed and marketed yearly. New development "Kojah" mink, so-called "Sable mink" longer-haired mutation introduced in late 1960's. Marketed in brown mink shades.

A new development, a recessive mink with fur two to three times as deep as regular mink, and apparently bearing a close resemblance to sable announced in November 1970, to be marketed in 1972. Announced by Genetics Department of University of Wisconsin. Dorsal sub-dermal fat, refined into light-colored oil now being sold as cosmetic base. Emollient qualities on human skin observed on handlers of mink carcasses. "Gigas" (giant) mink— 30″ from nose to rump. First offered for sale in commercial quantities December, 1970. This strain will be offered in sapphire and violet shades. Gigas is Greek word for "giant."

Wild U.S. Catch 1967–8: 183,625. From most states. Minnesota and Louisiana leaders. Total world production estimate 25,000,000 1970–71, ranch and wild.

First Mink Mutation: Ranch Mink

Early '70's Mutation: Jaguar Mink

Name: Mole

Size and Appearance: One of smallest animals used for fur. Working area approximately 3x5″. Natural color taupe gray. Very short fine hair, leather brittle around edges after dressing and dyeing. Fur has velvety look and feel.

Geographical Sources: Dutch and Scottish most common, also Belgium and Denmark.

Processing Used: Natural tipped, also dyed to all colors, easy to dye.

Durability: 20–10%.

Skin Layouts Used: Skins joined with rounded head-to-rump joining. Made into plate bodies, usually with hair direction reversed in each row. Horizontal and vertical.

Used For: Coats, jackets, costumes. Drapes like cloth and used like cloth for special effects.

Repairability: Poor. Best to remove skin if replaceable. Not always possible. Edge repair limited because of short hair.

Remodelling: Questionable unless almost new. Leather not strong. Has tendency to dry out.

Care: Not for daily wear. Avoid excess friction.

Look Alikes: Flat, sheared rabbit. Fabric imitations common.

Comments; History: Queen Alexandria of England had first garment made up of Scottish moles. She was successful in her objective of starting a fashion trend which would relieve Scottish farmers of a pest. Not popular in recent years but shows signs of returning to favor in the 1970's.

Name: Monkey

Size and Appearance: Size of monkeys in the zoo. Hair length varies from one to nine inches, depending on section and variety. Colors black, black and white, gray, gray and white, gray and red and vari-colored. Length 18″ to 24″ some to 30″. Silkiness most desirable. Has distinctive white mane from shoulders along sides to white-tipped tail. Leather not pliable.

Geographical Sources: Gold Coast, British West Africa, Abyssinia, central and east Africa, also Ethiopia.

Processing Used: Dressed and dyed black. Attempts to bleach and dye leather not too successful. Some black and white used natural.

Durability: 30–20%—most types.

Skin Layouts Used: Hair down to take advantage of sleekness. Often worked in strips as fringe, or alternated with ribbon or grosgrain.

Used For: More popular as edging and trimming on dresses and hats—also for jackets, capes, in rare fashion cycles.

Repairability: Easy to fix if enough fur. Damaged areas can be removed. Patching not difficult if fur is available.

Remodelling: Most garments made with alternating strips of cloth (grosgrain) and fur, thus complicating remodelling. All fur garments of monkey easy to remodel if leather in good condition.

Care: Keep fur brushed down in natural direction.

Look Alikes: Dyed long-haired goat and lamb, if straight and shiny.

Comments; History: Not fashionable for three decades. Some recent interest shown, but supply limited.

Name: Muskrat

Size and Appearance: Vest-pocket edition of beaver. Pelt works out to maltese cross shape, with point at tail. About 15″ long and same width. Natural dark-brown back, golden-brown sideback, silver flanks. "Jersey" muskrat almost black. Canadian and Great Lakes muskrats larger, more reddish, thinner leather.

Geographical Sources: U.S., two main sections, "Southern" Louisiana, Texas, "Northern" around Great Lakes. Russian variety smaller, very reddish, poor leather.

Processing Used: Dressing leather thickness varies with use. Skin and parts thereof extensively dyed and blended in many shades. Bellies sometimes sheared and grooved. Striping to imitate mink common. (Also see comments below.)

Durability: 70–60%, blacks (U.S.). 50–40%, flanks (U.S.).

Skin Layouts Used: Extremely varied. Skin worked natural in six-sided pattern effect. Backs zig-zagged together and striped, with and without dye to imitate mink. Bellies worked separately, zig-zag or pattern joining, dyed or natural. Skin also cut into three sections, back, gold (side back) and silver (flanks). Each worked separately in separate garments or in sequence of light to dark lines in ombre effect. Fine skins also let-out.

Used For: Perennial European favorite for linings, trimmings. Used for coats, jackets, etc., usually for casual garments.

Repairability: Excellent. Will accept edging well. Patching not too difficult if material available.

Remodelling: Bold patterns require usual careful handling. Leather has tendency to dry up with age and disintegrate when dampened. Avoid expensive remodels on older fur—shorten or make smaller.

Muskrat (rust-dyed)

Care: From above proper storage very important. Otherwise normal care.

Look Alikes: Dyed and striped marmot, rabbit.

Comments; History: Gets name from odor he emits (musk). Looks like rat. Once a pest, because so numerous. Northern varieties, once plucked, sheared and dyed black to make "Hudson Seal." In 1920, one of the most popular furs, now not made at all. Outstanding fur in demand in 1941. All phases of this fur extremely vulnerable to fashion changes. Introduced live to Germany in 1906, now a pest there. Russia introduced Canadian stock to Ural Region in 1926. Now (1968) second in world production.

Wild U.S. Catch 1967–8: 4,157,288. Most states. Louisiana, Ohio and Wisconsin the leaders.

Name: Nutria

Size and Appearance: First cousin to North American beaver. Naturally sepia brown underfiber with longer, lighter brown, heavy glossy guard hair. About 20″ long. Dressed skin forms a trapezoid shape, with rump wider than head.

Geographical Sources: Originally wild from Argentina, Brazil, Uruguay, and Chile. Some also farm-bred there. First imported for ranching to California in 1954. Also ranched in other countries. Also released and allowed to live wild in wetlands, especially in Louisiana.

Processing Used: Traditionally dressed plucked and sheared. Sometimes dyed black, brown and beige shades. Use of natural unplucked skin is new. Used unplucked and dyed black.

Durability: 60–50%.

Skin Layouts Used: Skin opened on grotzen to preserve belly, which is softest, silkiest and most useful part. Heavy leathered grotzen is trimmed off. Worked skin-on-skin with zig-zag. Fine sheared skins also let-out.

Used For: In plucked sheared form, traditionally used for luxury garments, also in dyed sheared form. Unplucked natural skins now used for casual men's and women's coats, parkas, linings.

Nutria (dyed)

Repairability: Excellent in both forms.

Remodelling: Leather is spongy and has tendency to weaken with age. Fine nutria can be remodelled with care if in good condition.

Care: Sheared types mat easily and require attention. Storing important. Unplucked form needs less care.

Look Alikes: In sheared form, beaver, nutria, mouton, processed rabbits. Unplucked, Canadian otter, fisher, natural raccoon.

Comments; History: Cuttings now being assembled into plates. "Nutria," Spanish for otter. Some attempts to raise under controlled herd conditions in Europe, North and South America, usually in semi-wild wetlands. Now so numerous in state of Washington they are classed as pest. Introduced to Israel where they helped clean out swamps and became a natural resource.

Wild U.S. Catch 1967–8: 1,120,942. Almost all from Louisiana, where it is competing with muskrat for swamplands.

Name: Ocelot

Size and Appearance: Smaller than leopard up to 30″ in length. Best background color is blue-gray. May be tan, gray or a rusty tan. Most attractive pattern of all cats, ranging from narrow black oval filled grotzen to almost all black. Ovals elongated, parallel to grotzen, angling away further from center back. Shoulder line at right angles to grotzen, accenting design.

Geographical Sources: Originally in southwest U.S., Mexico, most of northern and central South America. Mexican skins most desirable.

Processing Used: Natural, dressed.

Durability: 80–70%.

Skin Layouts Used: Skin-on-skin joining may be rounded head or elaborate hidden seam. Coat may need 1½ to 2 skins in length. Worked horizontal occasionally. Mostly hair down.

Used For: Fine coats and jackets.

Repairability: Difficult as for all boldly pattern furs.

Remodelling: Difficult, but warranted because of value of fur. Additional skins can be worked in if needed.

Care: Normal. With care, fur will last.

Look Alikes: Ocelocat, northern South America and Mexico. Size of house cat. Markings similar. Also prints on mink, calf, kidskins, lamb, etc. "Peludo"—same locality, size same as ocelot, marking, mixed tiger and ocelot.

Name: Opossum

Size and Appearance: American; head like rat, fur coarse, thick, underfur of gray and yellow, brown cast. Australian, more silky, dense, deep blue-gray. Tasmanian, red-brown. All about the size of a cat. Ringtail opossum has soft, short fur, light blue-gray, to dark brown, with white belly, smaller than above.

Geographical Sources: Six basic varieties: American, Australian, Tasmanian, Ringtail, South American and water opossum. Only first three important commercially. All except American from Australia.

Processing Used: Dressed. Clear blue and gray skins worked natural. Otherwise dyed and tinted in many shades. Good color, American skins also worked natural, otherwise dyed black or mink shades. Once sheared to imitate beaver.

Durability: 80–70% Australian. 50–40% U.S. 40–30% Tasmania.

184

Skin Layouts Used: Fine Australians and American formerly let-out, now only on fine Australians. Most dyed and blended and worked horizontal, also vertical, skin-on-skin.

Used For: Australian—Natural and tinted for fine casual and formal coats, jackets, capes. American—Natural split and full skin, horizontal and vertical, mostly coats and jackets.

Repairability: Normal.

Remodelling: Not warranted on worn American opossum. Acceptable on Australian if in excellent condition.

Care: Australian types mat easily and have tendency to develop wave. Requires periodic attention.

Look Alikes: Long-haired gray rabbits, dyed to match. Others in marsupial family include koala, wombat. Dyed long-haired rabbit resembles dyed opossum.

Comments; History: Not native to New Zealand. Introduced there about 1860. Zoologists list six different sub-species in New Zealand alone from silver-gray to black with chestnut red tinge. The opossums of Australia are not true opossums. "Phalanges" only native to Australia and Tasmania. Australian bears one or two young a year, the American sometimes bears twice a year—6 to 16 young.

Wild U.S. Catch 1967–8: 98,514. Missouri and Ohio leaders.

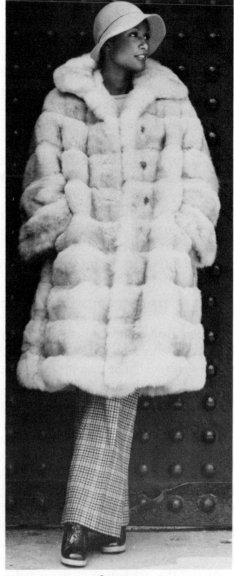

Opossum

Name: Otter

Size and Appearance: Several varieties marketed. Rarest sea otter. Very large, five to seven feet. Other varieties smaller. Fur rich, dense, lustrous, mahogany brown to light fawn color. Soft, long guard hairs. Body long and slender. Up to 40″ in length, except for sea otter and Brazilian which attain 60″ and more.

Geographical Sources: Sea otter, Arctic Canada. Canadian otter most of North Canada. (Brazilian) large dark-brown, (Ariranha) with short hair, also Asiatic, smaller, short hair, not so shiny. Brazilian otter larger, next to sea otter in size.

Processing Used: Fine dressing. All species usually marketed plucked and sheared. Canadian otter used unplucked. Brazilian, Siamese, also dyed beige shades.

Durability: Sea otter and Canadian—100–90%. South American—80–70%. Asiatic—70–60%.

Skin Layouts Used: Horizontal and vertical, full and split skin. Otter from colder sections can be let-out if necessary.

Used For: Canadian skins popular for trimming in natural form. Sheared forms used for jackets and coats.

Repairability: Excellent, except for very flat sheared varieties.

Remodelling: No patching possible, as with Alaska seal. Measurements of new pattern must conform fairly closely to original measurements of skins.

Care: Minimal. One of the most durable of furs. Fur looks better with age.

Look Alikes: Nutria, seals and beaver, if dyed and sheared to imitate.

Comments; History: Sea otter almost extinct. Carefully managed by Canadian government in 20th Century. First sale since 1911 Treaty in 1968, of 1,000 skins. Herd has been increasing due to careful conservation management.

Wild U.S. Catch 1967–8: 20,320. Florida, Alaska and Louisiana leaders.

Name: Pahmi

Size and Appearance: Small edition of badger, body only about 15″ long. Head mask like badger. Guard hair much darker than light underfiber, which is whitish, while top hair is brown to silver-gray. Lean, narrow body like ferret family.

Geographical Sources: China and India.

Processing Used: Originally plucked to show soft underfur, gray grotzen tone. Now being used natural and with slight tipping. Also plucked leaving dense soft underfur.

Durability: 60–50%.

Skin Layouts Used: Mostly rectangular, block form, horizontal and vertical. Skins are split and the left half-skins are matched together, and the right half-skins are sewed together.

Used For: Jackets and coats, womens' and mens' trimmings.

Repairability: Good.

Remodelling: Limited by total cost. Avoid elaborate remodel.

Care: Plucked forms may mat and rub. Requires regular care.

Look Alikes: Natural—flat foxes, dyed rabbits, dyed muskrat, female mink, susliki, peschaniki.

Comments; History: Known zoologically as "Ferret-Badger."

Name: Pony

Size and Appearance: Wild pony, not too dissimilar from pony used for children. Best grades show flowered moiré design near rump area. Mostly brown, occasionally brown-gray. Main character of pony skin focuses on two "flower" areas two-thirds down from head.

Geographical Sources: Russian and Polish. Some from Siberia, Mongolia, Denmark, South America. Best from Volga area, a few from Iceland, skins have long, wooly hair.

Processing Used: Normal dressing, usually dyed in dark-brown and black because of difficulty in matching out approx. 6 skins needed to make a coat.

Durability: 50–40%.

Skin Layouts Used: Hair down in most instances—moiré pattern on rump.

Used For: Casual jackets and coats.

Repairability: Poor, especially if edges are worn. Binding advised rather than cut away repair.

Remodelling: Pattern of skin may be deterrent. Addition of material will make job too costly for wear received.

Care: Vulnerable to hair breakage due to friction if rubbed against hair flow.

Look Alikes: Calf, kidskin, Lakoda seal.

Comments; History: Rise and decline in popularity very sharp.

Name: Rabbit

Size and Appearance: Varies in size and color from one commonly seen to "Giants" twice that size. Natural color usually brown, brown and white, but wild and cultivated colors range from white to almost black.

Geographical Sources: Australia, Belgium, Germany, France, New Zealand, Japan, China and South America, United States.

Processing Used: Complete range of all processing used on rabbits, dyed, plucked, sheared, grooved, printed to imitate any valuable fur. Also used natural in gray and white, also brown. Italians dye beautiful imitation chinchilla.

Durability: 20–10%, except for the fine-sheared, dyed French rabbits.

Skin Layouts Used: Block, horizontal, diagonal and vertical, split and full skins. Dyed skins sometimes zig-zagged to conceal joining.

Used For: The great imitator. Anything ever made in any other fur usually also made in rabbit. Also used for linings.

Repairability: Excellent. Can be re-edged easily in most instances.

Remodelling: Only salient factor is cost versus use after remodel. Usually not warranted except for fine sheared skins, because of high labor costs.

Care: Some natural, poorer skins wear away at edges after one full season of wear.

Rabbit (natural)

Others last three-five years. Normal care required. Almost any fur can be imitated in rabbit by proper processing.

Comments; History: Introduced to Australia 100 years ago. Northern third of continent is overrun. Has history of importation and migration over most of world.

Main drawback to once immense popularity is F.T.C. regulations requiring that rabbit product be so billed instead of misleading, but more glamorous "sealine," "beaverette," "chinchillette," etc. Fur sheds a great deal. There seems to be a problem of allergic reaction to rabbit with some people.

Name: Raccoon, American

Size and Appearance: Long-haired dense brown underfiber, with gray guard hair varying to black on center back. Size of medium dog. Tail color alternating black and tan rings. Flanks silvery. Best skins bluest, darkest on grotzen, absence of redness. Sweet water raccoon bluer, salt water more reddish.

Geographical Sources: Most United States, north central states best, especially Dakotas; Missouri, Arkansas and North Carolina next best. Southern raccoon mostly reddish.

Processing Used: Dressed and used in natural long-hair form, also semi-bleached. When sheared shows even dark dense pile, much like beaver. Sheared forms used natural and dyed, light tan and beige shades. Poor colored skins blended.

Durability: 80–70%.

Skin Layouts Used: Skin-on-skin with "V" head-rump joinings. Also let-out both long-hair and sheared. Bleached long-hair worked horizontal to imitate fitch, also long-hair natural skins.

Used For: Natural, for trimmings, jackets, coats, same for bleached long-hair. Sheared, coats and jackets, natural and dyed.

Repairability: Excellent. Accepts re-edging well on all phases.

Remodelling: Excellent on long-haired types if not let-out. Sheared and let-out

Raccoon

types more difficult. Leather is weaker if skin has been bleached.

Care: In sheared form, will mat. Needs periodic ironing to restore fluff.

Look Alikes: Asiatic and "Amur" raccoon, which have very long heavy reddish hair, used for trimmings. American ringtail, smaller, softer nap.

Comments; History: Has periodically re-emerged in fashion. In long-hair or sheared, popular in 1920 as "college" style. Popular in 1940's as sheared fur; resurgence

of long hair popular in last decade usually in form of hats and trimmings.

U.S. Raccoon commercially classed in three main types in long-hair form "coat" type. Shorter hair usually southern skins. Semi-heavy—medium height and weight, useful for coats and trimmings. Heavy—northern states, best for trimmings. Not applicable to sheared raccoon where density is most desirable characteristic.

Wild U.S. Catch 1967–8: 878,234. Ohio, Iowa, Missouri leaders.

Name: Sable

Size and Appearance: Soft skin, deep blue cast. Best with slightly silvered top hair. Hair very dense, turns equally well in all directions. Smaller working area than mink.

Geographical Sources: Siberian best. Bargvzin section finest ("Crown Sables"). Canadian sable not as dense or silky. Japanese and Chinese inferior. In 1970 Bargvzin ranched skins highest priced.

Processing Used: Dressed natural. Finest used in natural shade. Many tipped or blended to simulate best color. Once dyed blonde or beige shades.

Durability: 60–50%.

Skin Layouts Used: Full skin. Joinings head-to-head, rump-to-rump. Also let-out in fine garments. Worked horizontal and vertical.

Used For: Boas, scarves, trimmings, shrugs, throws, capes, jackets and some coats.

Repairability: Excellent. Edge repairs well.

Remodelling: Always worthwhile if fur and leather is in good condition. Can be re-blended if color has faded.

Care: Needs blowing out to keep from matting.

Sable

Look Alikes: American fisher (see Martens). Chinese sables, dyed martens.

Comments; History: Best qualities probably most valuable fur per square inch. Siberia explored by sable trappers. Supply decimated. U.S.S.R. now has preserves and controls harvesting. In August 1970 it was reported that "Soviets have succeeded in breeding black sable," after forty years of experiment.[4] Regal even in wildlife hierarchy. "Lesser" animals, swimmers ferry it across streams on their backs.

FUR SEALS

The glossary of seals in the bibliography lists 93 different seals, some used as fur, in existence today. Obviously not all of these are common enough to be discussed or described here. Only those which are significant, or have been a factor in the industry in the past, are considered. This does not preclude the possibility of one of them becoming fashionable or important in the future. Among those occasionally marketed are:

Rock Seal: South American sea lion, also known as "tropical seal." Moiré patterned flat seal.

Name: Fur Seal, Alaska

Size and Appearance: Dressed skin oval with flipper holes. Average length of dressed skin 40", plus or minus. Shape is oval with flattened edges. Head flatter than rump. Natural skin heavily protected by guard hair and dense underfiber.

Geographical Sources: Pribilof and nearby Bering Sea Islands of north Pacific area.

Processing Used: Processing of U.S. skins under government control by contract. Present contractor Fouke Fur Co. One of best processed furs involving 125 steps. Dyed black, plum ("Kitovi") dark-brown ("Matara") female (Lakoda) skins close sheared to about ⅛" used in natural russet brown, also dyed black and brown shades. All types plucked and sheared.

Durability: 100–90%.

Skin Layouts Used: Full skin, rump up (reverse) common on sheared furs.

Used For: Coats, some jackets. Occasionally fur trimmings.

Alaska Fur Seal (silver fox trim)

Repairability: Excellent. Can be re-edged and repaired.

Remodelling: Traditional difficulty with remodels which may involve adding width across shoulders or through top front. Skins cannot be patched or stretched in the width in any prominent areas. Fur amenable to re-working otherwise. Color can be re-done.

Care: Regular cleaning and ironing.

Look Alikes: Cape of Good Hope (African) duller, smaller, not as dense. Other fur seals also Uruguay seal, nicknamed "Lobos," now discontinued "Hudson Seal."

Comments; History: Mystery of where seal herd spends winter not completely solved. Comes to Pribilof Islands only for mating. Lakoda introduced in 1964—now also dyed "Sandrift" (light-brown), dark-blue, black and "coco" (1967). Only the young males (bachelors) are harvested each year. Through careful, government controlled harvesting, the herd has increased from 200,000 to about 1,400,000. This is close to the beneficial maximum for these animals.

Wild U.S. Catch 1967–8: 58,532. Alaska.

Name: Seals, Hair

Size and Appearance: 1. Saddler (young harp seal) seal, short-haired fur with irregular blue or brown spots on a silver-gray background. So called "white coat" is pup of harp or saddleback up to ten days old. Distinguished by white flanks. By 21 days changes to "Beater" light-gray, scattered medium gray spots. Fully moulted. Beater stays in this phase until about 60 days in age, when it develops characteristic saddle shape design on back. Misnamed "Greenland" seal.

2. Ring seal—pattern of light rings in dark layer, lighter sides.

3. Hooded (blueback) seal until six months old. Blue back and contrasting white sides.

4. Ranger seal—very dark, medium dark sides, lighter small spots, velvety look.

Geographical Sources: Most seals in Arctic areas of the world, mostly North American and adjoining waters. Many from St. Lawrence Gulf and seas off Labrador and Newfoundland.

Processing Used: Stained yellowish skins bleached if to be used naturally.

Hair Seal

Often dyed black and dark-brown shades.

Durability: 70–60%.

Skin Layouts Used: Some hair up, mostly hair down. Occasionally worked natural split skin horizontally.

Used For: Coats, jackets, ski-parkas.

Repairability: Fair. Edging limited by prominence of pattern. Dyed types easier to repair.

Remodelling: Limitations of skin design and size to new pattern prevent radical style changes without adding full skins. Match difficult—may not be warranted with average worn coat.

Care: Most hair seals have stiff hair, prone to wear. Daily wear not advised. Normal care.

Look Alikes: In dyed forms, often difficult to distinguish from one another. Constant handling by specialist, often only way to discover proper designation, among many other rarer varieties not listed above.

Comments; History: In 1970 Canada secured agreement to eliminate killing of whitecoats, limited killing at "beater" stage of growth. As late as 1964 Canadian and Alaskan Governments paid $5.00 bounty for each hair seal caught off Pacific Coast.

Name: Skunk

Size and Appearance: Size of small dog. Naturally blackish brown with "V" shaped white marking of varying length on back. Best almost blue-brown, glossy and dense, rather than brownish.

Geographical Sources: North American skunk. Best from Dakotas and Minnesota. Also some from western Canada, South American-zorrina, smaller, softer leather and hair.

Processing Used: Best grades dressed and used in natural color. Others dyed in brown shades and black, after white stripe is cut out. Some of good quality and poor color formerly tipped to improve color.

Durability: 80–70% — U.S. 50–40% — Zorrinas.

Skin Layouts Used: In 1930–40's worked let-out like mink, with white stripes removed, joined and used as trimmings (skunk-whites). In 1967–68, coats made skin-on-skin with white marking left in,

horizontal split skin and vertical.

Used For: Trimmings, parkas, jackets and coats. More popular in Europe now. Made into coats in U.S. in 1940's.

Repairability: Excellent. Can be re-edged without any trouble.

Remodelling: Rare problem in these days. Let-out garments not made. Not too difficult on full skin garments, with pattern. Cost versus utility must be compared.

Care: Normal. Fur develops slight musty smell when wet, disappears when dry. May be objectionable. No residue of skunk odor on dressed skin.

Look Alikes: Civet cat, little striped skunk. Zorrina, South American skunk smaller.

Comments; History: Ranching attempted when fur was popular (scent glands removed).

Wild U.S. Catch 1967–8: Skunk and Civet cat 19,459.

Name: Squirrel, Flying

Size and Appearance: Mottled mouse color. Gray tinged with black; reddish when unprime, long silky hair, 18″x6″ average

working area. Leather thin and soft.

Geographical Sources: China and East Siberia.

Processing Used: Dressed—very often dyed to imitate more expensive long-haired foxes.

Durability: Poor.

Skin Layouts Used: Full and split skin.

Used For: Trimmings. Once made into yardage edging for dresses, etc., when such styles were popular.

Repairability: Poor.

Remodelling: Not advised.

Care: Not for any use involving friction or daily wearing.

Look Alikes: Small dyed foxes, long-haired. Skunk, etc., if dyed.

Comments; History: Has fluttery membrane between front and back legs along sides. Jumps and "sails" from limb to limb.

Name: Squirrel

Size and Appearance: Siberian and Canadian about size of kitten. Thin tapering body. Rump twice as wide as head. Best grades dense, blue-gray color, no discoloring, lower center back streak. Poor grades reddish. Belly white or off-white, good indicator of grade of skin. Fiber one of the softest of all commercial furs, leather one of most elastic.

Geographical Sources: Asia, Europe and North America. Best from Siberia. Canadian skins less dense, useful if dyed. U.S. squirrels not useful as fur.

Processing Used: Only best clear blue-grays used in natural color. Rest dyed in various light to dark-brown shades, usually to imitate mink. Bellies worked separately into plates ("sacques" for squirrel) in natural white with gray edges, or dyed into various shades.

Durability: 40–30%.

Skin Layouts Used: Almost exclusively rounded or "V" head-to-rump joining. Sometimes pre-assembled into 48 or 50 skin plates. Bellies always pre-assembled into plates, sometimes diagonal and chevron layouts. Sometimes joined with zig-zags.

Used For: Once a popular coat and jacket item in natural or dyed shades. Dyed squirrels used extensively in stoles, clutch capes, etc. to imitate mink. Bellies popular for baby carriage robes, lining children's

Squirrel (sides)

clothes. Care must be exercised to avoid possible allergic reactions.

Repairability: Edges and fur in general not difficult to repair.

Remodelling: Easy to reshape because of elasticity of the leather. Not too difficult to add fur if needed. Only best natural grades warrant extensive remodel.

Care: Because of extreme softness, fiber will wear if it becomes matted. Must be kept fluffy to prevent wear at fiber-base from friction.

Look Alikes: Some rabbits. Canadian for Russian if dyed.

Comments; History: One of many furs once reserved for royalty. Old-fashioned name for squirrel skin plate was "sac" or "sacque." In Colonial days was killed for a bounty because it was considered a pest.

Name: Suslik

Size and Appearance: Natural yellowish gray, with shiny silky guard hair. Smallish size of a kitten. Peschaniki, one form of suslik.

Geographical Sources: Mainly Russia.

Processing Used: Usually dyed in mink and sable shades. Lately being worked natural.

Durability: 30–20%.

Skin Layouts Used: Split skin in horizontal, chevron and vertical patterns. Hair down.

Used For: Jackets and coats. Some cloth coat trimmings in dyed forms.

Repairability: Fair.

Remodelling: Probably inadvisable except if labor costs minimal.

Care: Not durable. Must be worn accordingly.

Look Alikes: Fine dyed garments resemble weasel and female mink. Difficult to identify in dyed forms.

Comments; History: Misnamed "Russian sand weasel."

Name: Viscacha

Size and Appearance: Small rat-like creature. Natural color ranges from tan white flanks and a mottled gray grotzen which shades off.

Geographical Sources: Argentina and Patagonia.

Processing Used: Dressed, dyed and usually striped to imitate costlier furs like mink.

Durability: 20–10%.

Skin Layouts Used: Full skin. Zig-zag or plain joinings if made into plates for dyeing.

Used For: Popular-priced trimmings and novelties, when used. Not seen much in U.S. in 1960's.

Repairability: Poor.

Remodelling: Not advised.

Care: Not for everyday wear.

Look Alikes: Marmot, squirrel, pahmi, rabbit, especially when dyed.

Comments; History: Gopher-like living environment and like gopher once regarded as pest by localities it infested. This burrowing creature lives in communities of up to 30 with interconnected burrows.

Name: Wallaby

Size and Appearance: Fur wallaby—red to dark-gray dense full fiber. Rock wallaby—grayish with distinct reddish central back-line, flatter and shorter hair than wallaby.

Geographical Sources: Australian fur.

Processing Used: Dressed, blended to enhance gray tone if possible, or dyed.

Durability: 40–30%.

Skin Layouts Used: Full and split skin layouts.

Used For: In 1970, some use for skin parkas. In general, a novelty fur.

Repairability: Fair.

Remodelling: Probably not advisable. Cost of remodel too great for investment return, unless practically new. Additional fur not readily available.

Care: Normal. Will mat and curl. Needs regular care.

Look Alikes: Opossum, squirrel, Koala bear, small kangaroo, bassarisk, all bear. Some resemblance if all are dyed. Looks somewhat like beaver if sheared.

Comments; History: A member of the marsupial family.

Name: Weasel

Size and Appearance: Small, long narrow body, "long rat." Color varies with localities and with seasons in cold countries. Short-tailed, good top hair and underfur. Southern varieties remain brown all winter. Basic color creamy or yellow, poor grades more brownish. Working length to 12 inches. American fur weasels have tipped tail.

Geographical Sources: Asia, Siberia, Japan, China, North America, Alaska, Canada.

Processing Used: Mostly dyed.

Durability: 40–30%.

Skin Layouts Used: Usually full skin with "V" or round-head joining. Also split skin chevron. Very fine full skins let-out.

Used For: Trimmings, jackets, some coats, occasionally cloth coat lining.

Repairability: Fair—can be re-edged.

Remodelling: Not practical except for fine grades of the skin.

Care: Probably not useful for daily wear. Normal attention required.

Look Alikes: Ermine, denser fur, long-tailed, black-tipped. Year round brown weasel, misnamed "Summer Ermine."

Comments; History: Biological classification "Weasel" includes ermine, mink, marten, skunk, otter and many other animals. American short-tailed ("least") weasel not useable as fur.

Wild U.S. Catch 1967–8: 6,997. Alaska and Michigan.

Name: Wolf

Size and Appearance: 1. Timber or gray wolf; light gray, blue-gray, brown and black.

2. Siberian; paler gray and coarser fur.

3. Russian black wolf; long black guard hairs. Most wolves long, up to five feet and narrow in width. Best are white, gray tones, gray-black and blue.

4. Prairie wolf; best pale gray, brown, gray-silver. Poor skins have tawny, doggy, or reddish.

Geographical Sources: Russia, Siberia, North America including Mexico. Latter has silkiest dense fur. Best American grades from Arctic area and Alaska.

Processing Used: Dressed. Off-color skins bleached and dyed to imitate lynx.

Durability: Timber to 60%, Prairie 40%.

Skin Layouts Used: Full and split skin. Center grotzen (mane), consisting of flat coarse hair cut away in use. Can be let-out.

Used For: Mostly as a trimming, although recently popular for fine ski parkas.

Rarely used today because of limited supply.

Repairability: Good—can be re-edged.

Remodelling: If in garment form, may be warranted.

Care: Normal attention and care.

Look Alikes: Coyote, is size of medium dog, reddish gray, tipped with black. Brush wolf from Ontario, Manitoba and Central U.S., has short even fur 1½″ to 2″. Top hair is blackish, sometimes silvery, underfur yellowish. Also lynx and lynx cat, also Asiatic dogs.

Comments; History: Wolves crossing into Arizona from Mexico killed by U.S. Fisheries and wildlife agents to prevent killing of cattle in 1970. In 1970 total estimate wolf population North America including Canada, 50,000. Almost extinct in Scandinavia and Japan, declining in Mexico. Extensively hunted and poisoned in U.S. Number has declined tremendously to near extinction.

Wild U.S. Catch 1967–8: Timber 1,709. Almost all from Alaska.

Name: Wolverine

Size and Appearance: When live looks like miniature bear, with underfiber dark brown set off by whitish stripe from shoulders along sides coming together at tail base. Body about 2 to 2½ feet from nose to tail. Has long grotzen hair and dense under fiber.

Geographical Sources: Northwestern border area, Canada and U.S. Also through cold belt of Europe and Russia.

Processing Used: Dressed. Not normally dyed, but may be blended.

Durability: 100–90%.

Skin Layouts Used: Usually worked full-skin, hair down to set off stripe ovals. Also as trim as indicated.

Used For: Special use as face edging on cold-weather garments for Eskimos. See comments below.

Repairability: Excellent—re-edges well.

Remodelling: Warranted if in garment form. Fur durable and can be worn for long time.

Care: Normal.

Look-Alikes: Small bear sections, wolf, large skunk.

Comments; History: In strong demand during World War II. Fur best for protecting face, does not attract frost. Used for this purpose by Eskimos.

Wild U.S. Catch 1967–8: 270. All from Alaska.

Name: Zebra

Size and Appearance: Size and shape of small, plump horse. Striping as commonly seen on zoo animals. Skin nearly as wide as it is long.

Geographical Sources: Sudan to Ethiopia.

Processing Used: Natural dressing.

Durability: Not good. Hair too stiff, leather strong.

Skin Layouts Used: Full skin only.

Used For: Coats and floor rugs, upholstery.

Repairability: Almost impossible. Pattern too strong.

Remodelling: Fur too new for any reliable information. Not possible to patch.

Care: Avoid friction.

Look Alikes: Calf, kidskin, mink, mink piece prints.

Author's Note . . .

The best and most exhaustive study on fur pelts, skins and hides as such, their appearance, origin and uses was compiled by Arthur Samet in his *Encyclopedia of Furs* (see bibliography). Written in 1950, it remains the best work of its kind, despite the inevitable obsolescence of one or two items (the same is true of some manufacturing techniques which appear in this author's *Fur Book* also written during the same period). Acknowledgement is freely made to frequent use of Samet's book to verify some parts of the preceding fur descriptions.

FOOTNOTES—CHAPTER 11

1 Dr. Paul Scops, in cooperation with Dr. H. Brawkhoff, Dir. Ing. A. Ginzel, Kurt Hase, and Richard Konig, "The Durability of Fur," *Berlindas Pelzgewerbe*, no. 445, 1963, trans. by Francis Weiss.
2 "Trappers Take Increased in '69," *Women's Wear Daily*, Nov. 9, 1970.
3 "Shearling Look Makes Ski Scene," *Women's Wear Daily*, Nov. 24, 1970, p. 33.
4 Jury Frankel, "Sables," *International Fur Review*, London, May, 1969, p. 52.

APPENDIX

APPENDIX

Appendix A. Endangered Species Act of 1969

The Endangered Species Confiscation Act of 1969 was enacted to "govern the importation and transportation of endangered wildlife." Included in the regulations is an "Endangered Species List" containing "wildlife found in other countries that are threatened with world-wide extinction" and the same "native to the United States."[1] The regulations describe the procedure by which all imports of wildlife shall be made, involving certain ports of entry with required authorizations and declarations.

The heart of the act is the "United States list of endangered wildlife." The list is in three columns. The first contains the common name, the second lists the scientific name, and the third indicates where the animal is found. The following is a modification of that list, omitting the scientific name and adding (in parenthesis) an explanation of the common name if it is unusual. This list, extensive as it is, will undergo frequent changes as the ecological balance in any state or foreign country changes. It contains, at last printing in 1966, 14 U.S. mammals and 160 foreign mammals.

Endangered Mammals, Indigenous to U.S.:

Indiana Bat

Utah Prairie Dog

Delmara Peninsula Fox Squirrel

Eastern Timber Wolf

Texas Red Wolf

San Joaquin Kit Fox

Black Foot Ferret

Florida Panther

Caribbean Monk Seal

Guadaloupe Fur Seal

Florida Manatee (Florida Sea Cow)

Key Deer

Columbian White-Tailed Deer

Southern Pronghorn

Endangered Mammals, U.S. and Foreign Origin

Common Name	Where found	Common Name	Where found
Thylacine (wolflike marsupial)	Tasmania	Haitian solendon	Dominican Republic
Cuban solenodon (insect eater)	Cuba	Lemur—all species	Madagascar and Comoro Islands

Common Name	Where found	Common Name	Where found
Indris, Sifakas, Avahis	Madagascar and Comoro Islands	Orangutan	Indonesia, Malaysia, Brune
Aye Aye (nocturnal lemur)	Madagascar	Gorilla	Central & West Africa
Spider monkey (4 species)	Guatemala and Costa Rica	Brazilian three-toed sloth	Brazil
Red-backed squirrel monkey	Costa Rica	Pink fairy armadillo	Argentina
Wooly spider monkey	Brazil	Volcano rabbit	Mexico
		Mexican prairie dog	Brazil
White-nosed saki (sub species monkeys)	Brazil	Thin-spined porcupine	Brazil
		Sperm whale	world-wide
Vakari, all species (short-tailed South American monkey)	Peru, Colombia, Venezuela, Brazil and Ecuador	Baleen whales	world-wide
		Northern kit dog	Canada
Goeldis marmoset	Brazil	Asiatic wild dog	Russia, Pakistan, India, Central & S.E. Asia
Golden rumped, golden-headed Tamarin golden marmoset	Brazil		
		Mexican grizzly bear	Mexico
Lion-tailed macaque	India	Formosan yellow-throated marten	Formosa
Tana river mangabey (slender long-tail monkey)	Kenya	Black-footed ferret	U.S. & Canada
Duoe langor (Asiatic mangabey)	Indo China	Cameroon clawless otter	Cameroons
Pagi Island langur	Indonesia	La Plata otter	Uruguay, Argentina, Brazil
Red colobus (slender long-tailed monkey)	Kenya	Giant otter	Amazon Basin
		Barbary hyena	Morocco
Zanzibar red colobus	Tanzania	Brown hyena	Southern Africa
Kloss' gibbon	Indonesia	Asiatic cheetah	Russia, Afghanistan, Iran (formerly India & Pakistan)
Plieated gibbon	Malaysia		

Common Name	Where found	Common Name	Where found
Spanish lynx	Spain	Sumatran rhinoceros	S.E. Asia, E. Pakistan, Vietnam, Indonesia
Barbary serval (long-limbed African cat)	Algeria	Javan rhinoceros	Indonesia, Burma, Thailand
Formosan clouded leopard	Formosa	Northern white rhinoceros	Congo, Uganda, Sudan
Asiatic lion	India	Pygmy hog	India & Nepal
Sinai leopard	Sinai, Saudi Arabia	Vicuna	Peru & Bolivia
Barbary leopard	Morocco, Algeria, Tunisia	Swamp deer	India and Nepal
Anatolian leopard	Lebanon, Israel, Jordan, Turkey, Syria	Kashmirstag, hangul	Kashmir
		Barbary stag	Morocco, Tunisia, Algeria
Bali tiger	Bali	M'Neill's stag	China, Tibet
Javan tiger	Indonesia	Shou (red deer)	Tibet, Bhutan
Caspian tiger	Russia, Afghanistan, Iran	Brown antlered deer	India, S.E. Asia
Sumatran tiger	Indonesia	Persian fallow deer	Iraq, Iran
Mediterranean monk seal	Mediterranean	Bowean deer	Indonesia
		Marsh deer	Argentina, Uruguay, Brazil
West Indian (Florida) manatee	U.S., Costa Rica, Guatemala, Panama, Brazil, Venezuela	Sondran pronghorn	Mexico, U.S.
		Black-faced impala	S.W. Africa, Angola
Amazonian manatee	Peru, Amazon River basin	Swayne's hartebeest	Ethiopia
Asian wild ass	Ethiopia, Sudan, Somalia	Anda (wild dwarf buffalo)	Indonesia
Mountain tapir	Columbia, Ecuador	Tamaran (small wild buffalo)	Philippines
Brazilian tapir	Venezuela, Brazil, Argentina	Wood bison	Canada
Central American tapir	Guatemala, Costa Rica, South Mexico, Colombia, Ecuador	Seladang (Guar.) (wild buffalo)	India, S.E. Asia
		Wild yak	Tibet

Common Name	Where found	Common Name	Where found
Kouprey (wild ox)	Cambodia	Cavieri gazelle	Morocco, Tunisia
Banteng	S.E. Asia	Slender horned gazelle	Spain, Egypt, Algeria, Libya
Pyrenean ibex	Spain		
Walia ibex	Ethiopia	Black lechwe (antelope)	Zambia
Rio de Oro Dama gazelle	Spanish Sahara	Kafue lechwe	Zambia
Mhorr gazelle	Morocco	Arabian oryx (antelope)	Arabian Peninsula
Moroccan Dorcas gazelle	Morocco	Clark's gazelle, dibatag	Somalia, Ethiopia

Appendix B. State Regulations Affecting Animal Life

STATE LAWS AND REGULATIONS AFFECTING WILD ANIMAL LIFE

The U.S. Bureau of Wildlife and Fisheries supplied the names and addresses of appropriate state agencies to whom a simple questionnaire was sent. They were asked to name the wild fur-bearing animal's within their borders: [2]

1. Those animals under partial protection by controlled killing;

2. Those animals under full protection from any form of harm or destruction, as endangered species;

3. Those animals regarded as dangerous, too numerous, or predatory, for which extermination or thinning-out programs are encouraged.

The following chart is a compilation of the replies received. Most animals listed as "not protected" had no closed season as such, but the method of killing was usually strictly limited. Many state regulatory agencies simply stated that their general policy was to cooperate with the U.S. Bureau of Wildlife and Fisheries, without being specific on how it affected the fur-bearers in the state.

State	Fully Protected	Partially Protected	No Restrictions	Predatory or Bounty Paid
Alabama	Black bear	Civet cat, mink, muskrat, opossum, otter, raccoon, skunk, weasel, nutria	Beaver, fox, bobcat	None
Alaska	None	Sea otter, beaver, coyote, Arctic fox, red fox, lynx, marten, mink, weasel, muskrat, land otter, raccoon, red squirrel, flying squirrel, ground squirrel, wolf, wolverine	None	None
Arizona	Ocelot, jaguar, wolf, otter, mink, cottontail rabbit	Bear, squirrel, muskrat, badger, raccoon, beaver, bighorn sheep, elk	Opossum	Mountain lion, bighorn sheep

State	Fully Protected	Partially Protected	No Restrictions	Predatory or Bounty Paid
Arkansas	Red wolf, mountain lion, black bear	Otter, beaver, raccoon, opossum, civet cat, skunk, red and gray fox, weasel, muskrat	None	Bobcat, coyote, nutria
California	Fisher, marten, wolverine, river otter, kit fox, ringtail cat	Mink, gray fox, cross fox, silver fox, red fox, badger, muskrat, beaver, raccoon	Coyote, weasel, skunk, bobcat, opossum	None
Colorado	Fisher, river otter, lynx, pike, black-foot ferret, wolverine, Albert's squirrel, long-tail weasel, muskrat, ringtail cat	Badger, marmot, marten, short-tail weasel, cottontail rabbit, snowshoe hare, kit fox, beaver, mink, long-tail weasel, muskrat, ringtail cat	Raccoon, short tail gray fox, bobcat, white-tail rabbit, hognose skunk, opossum, coyote, red fox, striped and spotted skunk, black-tail rabbit	None
Connecticut	Gray squirrel, cottontail rabbit, snowshoe and European hare, red squirrel, woodchuck	Mink, muskrat, otter, raccoon, bobcat, Canada lynx, fisher, gray and red fox, opossum, weasel, skunk	Raccoon, muskrat, beaver	None
Delaware	Delmarva fox, squirrel	Muskrat, mink, river otter, opossum, raccoon, red and gray fox, chipmunk	None	None
Florida	Everglades mink, Florida weasel, panther, key deer, round-tail muskrat	Squirrel, deer, bear	Raccoon, opossum, red and gray fox, beaver, bobcat, skunk, flying squirrel, jaguarondi, nutria, mole	None
Georgia	Alligator	Mink, muskrat, opossum, skunk	None	None
Hawaii	None	None	None	None

State	Fully Protected	Partially Protected	No Restrictions	Predatory or Bounty Paid
Idaho	Fisher, wolverine	Beaver, marten, mink, muskrat, otter	Coyote, skunk civet cat, weasel, badger, red fox, nutria, raccoon, bobcat, lynx, jack rabbit	None
Illinois	Badger, otter	Opossum, raccoon, weasel, mink, muskrat, beaver	Skunk, red and gray fox	None
Indiana	Badger, bobcat, otter, black squirrel	Opossum, raccoon, weasel, mink, muskrat, beaver	Red and gray fox	None
Iowa	Otter	Beaver, badger, mink, muskrat, raccoon, civet cat, weasel, wolf, groundhog, red and gray fox	Coyote	None
Kansas	Otter, swift (prairie) fox	Beaver, mink, muskrat, opossum, weasel	Raccoon, skunk, red and gray fox, badger, bobcat, coyote	None
Louisiana	(alligators)	Nutria, muskrat, raccoon, beaver, mink, otter, opossum, skunk, spotted ringtail cat	None	Nutria (on productive agricultural land)
Maine	Marten, lynx	Otter, mink, muskrat, fisher, beaver, raccoon	Eastern coyote, fox, bobcat, skunk	Bobcat
Maryland	None	Otter, raccoon, opossum, beaver, muskrat, mink, skunk, red and gray fox	None	None
Massachusetts	Otter	Muskrat, mink, otter, beaver, raccoon, opossum	Weasel, red and gray fox, skunk, wildcat	None

State	Fully Protected	Partially Protected	No Restrictions	Predatory or Bounty Paid
Michigan	Pine marten, fisher (reintroduced), wolf, lynx, elk, moose	Deer, bear, cottontail rabbit, snowshoe rabbit, squirrel, bobcat, mink, raccoon, badger	Coyote, fox, opossum, weasel, red squirrel, skunk	Coyote
Minnesota	Fisher, pine marten	Otter, beaver, muskrat, mink, black bear	None	Red and gray fox, coyote, timber wolf, bobcat
Mississippi	None	None	None	None
Missouri	Raccoon, opossum, weasel, skunk, civet cat, badger, red and gray fox, bobcat, mink, muskrat, beaver	None	Coyote	None
Montana	Black-foot ferret, fisher	Mink, marten, beaver, muskrat	None	Fox, lynx, bobcat, coyote
Nebraska	None	Muskrat, mink, beaver, squirrel, cottontail rabbit, deer, antelope, coyote	None	Coyote, wolf, bobcat, badger, opossum, raccoon, skunk
Nevada	Mountain beaver, kit fox	Red and gray fox, marten, mink, muskrat, nutria, beaver, otter	Coyote, bobcat	None
New Hampshire	Marten, all seals	Otter, mink, muskrat, raccoon, fisher, beaver	Skunk, fox, weasel	Bobcat
New Jersey	None	Mink, muskrat, raccoon, fox, beaver, otter	None	None

State	Fully Protected	Partially Protected	No Restrictions	Predatory or Bounty Paid
New Mexico	Otter, black-foot ferret, pine marten, coatmandi, Alpine marmot, mink	Muskrat, beaver, weasel, nutria, fox, ringtail cat, raccoon, badger	None	Coyote, bobcat, skunk
New York	Marten	Fisher, raccoon, skunk, mink, otter, beaver, muskrat	Opossum, weasel, coyote, bobcat	Red and gray fox
North Carolina	Otter	Muskrat, raccoon, opossum, beaver, mink, squirrel, rabbit, fox	None	Skunk, weasel, wildcat
North Dakota	Wolf, beaver, fox, coyote	Mink, muskrat, weasel, coyote, red, gray and swift fox	Badger, skunk, raccoon	None
Ohio	None	Mink, raccoon, opossum, muskrat, beaver	Skunk, weasel	None
Oklahoma	Otter	Weasel, raccoon, beaver, muskrat, mink	None	Coyote, bobcat, skunk, opossum, civet cat
Oregon	Fisher, ringtail cat, wolverine, sea otter	Mink, marten, muskrat, beaver, river otter	Raccoon, coyote, red and gray fox, bobcat, weasel, skunk, civet cat, badger	Nutria, opossum
Pennsylvania	Otter, bobcat	Beaver, muskrat, mink	Red and gray fox, skunk, opossum, weasel	None
Rhode Island	None	Muskrat, mink, otter, skunk, raccoon	Opossum, mole, rabbit	None
South Carolina	None	Raccoon, opossum, fox, mink, skunk, otter	None	None

State	Fully Protected	Partially Protected	No Restrictions	Predatory or Bounty Paid
South Dakota	Black-foot ferret	Silver, black and cross fox, opossum, muskrat, otter, mink, beaver, weasel	None	Raccoon, skunk, red and gray fox, coyote
Tennessee	Otter (plus all listed by U.S. Dept. Wildlife and Fish)	Squirrel, rabbit, raccoon, opossum, mink, white-tail deer, bear, wild hog, red fox	Skunk, weasel, bobcat, gray fox, beaver, coyote	None
Texas	None	Beaver, otter, mink, ringtail cat, badger, skunk, raccoon, muskrat, bear	None	Wolf, coyote, prairie dog, ground squirrel, wildcat
Utah	None (Option open on otter, mink, marten, sable)	Cougar, bear, otter, mink, marten, sable, elk, antelope, deer	None	Coyote, bobcat
Vermont	Fisher, pine marten, caribou, elk, moose	Beaver, otter, mink, raccoon, skunk, red and gray fox, muskrat	Red and gray fox	Bobcat, lynx
Virginia	Elk	Bear, mink, muskrat, beaver, otter, deer, fox, raccoon, squirrel, opossum, civet cat	None	Weasel, wildcat, skunk
Washington	Fisher, wolverine	Beaver, bobcat, mountain or cross fox, Canada lynx, marten, muskrat, river otter, weasel, mountain lion	Coyote	Nutria
West Virginia	Otter	Mink, muskrat, fisher, raccoon, snowshoe hare, cottontail rabbit, beaver	Red and gray fox, skunk, opossum, woodchuck	None

State	Fully Protected	Partially Protected	No Restrictions	Predatory or Bounty Paid
Wisconsin	Marten, fisher, wolverine, badger, timber wolf, Canada lynx	Raccoon, wildcat, muskrat, mink, woodchuck, beaver, otter, black bear	Coyote, red and gray fox, opossum, skunk, weasel	None
Wyoming	Otter, fisher	Marten, mink, beaver	None	Coyote, bobcat

FOOTNOTES—Appendix A and B

1 Federal Register, Vol. 35, No. 106, Tuesday, Jun. 2, 1970, p. 8495.
2 Office of Conservation Education, Bureau of Sport Fisheries and Wildlife, U.S. Dept. of Interior, Washington, D.C. 20240 W.L. 435. R, Nov., 1969.
"State Wild Life Agencies" Office of Conservation Education, Bureau of Sport Fisheries and Wild Life. U.S. Department of Interior, Washington, D.C. 20240 W.L. 435R, Nov. 1969.

Appendix C. Answers to Key Questions People Ask About Furs

A COMPENDIUM OF QUESTIONS AND ANSWERS

Any experienced handler of furs or, for that matter, any knowledgeable customer, is likely to hear or ask certain questions over and over. A sampling of some of the most common and important follows.

Q. 1 Why should I buy a fur coat rather than a cloth coat?

It is not a matter of buying one rather than the other. Each has its place. The advantages in having fur is that fur is warmer and more prestigious. Fur usually lasts longer. Fur plays an indispensable part in the wardrobe of the fashion conscious woman.

Q. 2 What type of garment shall I buy?

What type you should buy depends on your wardrobe. If you can afford to buy a whole new outfit to go with your casual everyday new fur, fine. On the other hand, you may have specific formal occasions in mind. You must picture yourself in the places where you will wear your fur, the rest of the ensemble with which you plan to use it, and buy accordingly. The region in which you live is also a factor as is your individual reaction to cold. Some can walk around in freezing weather with a thin topcoat. Others must bundle up at the first snap of winter and may have to compromise with appearance and utility if they are on a strict budget. There are several moderate-priced or better furs which are both light and warm. The perennial popularity of mink and Alaska seal is in part because they fulfill both of these requirements, in addition to being beautiful. Most furs are warm, but not all are light in weight.

The average customer will probably find a coat or jacket a good beginning for a fur wardrobe. If you travel in winter, fall, and spring by car, the jacket may be enough over a gown even in cold weather for the short dash from curb to door. On the other hand, in localities where it is often below freezing for weeks at a time, a warm full length fur may be the best answer.

Q. 3 Which fur shall I buy?

The multiplicity of furs and their processed variations are so numerous as to make this question almost unanswerable. Is warmth a primary consideration? Any of the furs that are amphibious and are equipped by nature with a dense underfur or fiber to keep the animal warm in cold water will fill the bill. Beaver, nutria, mink, the fur seals, otter, and muskrat are some of the furs in this category. On the other hand, you may be more interested in how a fur looks on you rather than its warmth. Try the garment on, and be honest with yourself. Let's face it, a short, chubby, size 16, isn't going to look long and lean in a long-haired fox coat.

Q. 4   What help do I need in selecting a style?

Often the prospective buyer who isn't sure of her own taste and judgment will bring along a friend, relative, or spouse in whom he or she has confidence to help make the decision. Sometimes such experts can be of great help. Trouble arises when the expert is of another generation with different tastes or is too opinionated. The experienced furrier who rightfully regards every one of his garments as a walking advertisement for his business is far more likely to be of constructive help than these onlookers.

In selecting a given style keep in mind an oft-proven proverb: "The sharper a style, the sooner it will lose its edge." A good example was the Nehru Jacket, which in the last decade blossomed and died in six short months. Your choice should depend on how it looks on you, how long and how often you intend to use it, and whether it can be remodelled. Don't buy a style because you saw it on someone and it looked beautiful. Keep in mind the key issue: how it looks on you!

Q. 5   Why are furs so expensive?

They aren't! In comparison with fur prices two decades ago and the rising cost of everything else furs have risen little. In some instances they are in about the same price bracket that they were 20 years ago. Some fashionable and very serviceable fur coats cost the same or less than a fur-trimmed cloth coat, a situation unheard of 20 years ago.

Some wild peltries have become rare and over-priced but it is not true that the best furs are the most expensive. Rarity, cost of growing and processing, and the manufacturing costs all influence the final selling price. Sudden demands for a particular skin in short supply are unpredictable and so are their effects on the selling price.

On the other hand, consider a consistently low-priced fur like mouton (dyed sheared lamb). Over the last decade this very warm and serviceable fur has been marketed at a selling price about equal to, if not less than, a good cloth coat. For warmth, durability and ease of repair, this fur is the better buy.

For most furs, except the very inexpensive and the let-out garments, labor represents a small percentage of the cost of the garment. By far the most important is the cost of the fur itself. This changes not only from one species to another, but within the grades of the *sections* of the pelt itself. The latter refers to the geographical area from which the skin comes. For instance a fine Siberian squirrel is worth two to three times its Canadian brother just as a Minnesota raccoon sells for twice the price of its Louisiana cousin. Quality, not rarity is the deciding factor in these instances.

Q. 6   How often should a fur garment be cleaned?

The answer depends on the type of fur, the locality in which it is worn, and the number of wearings. A sheared fur such as Alaska seal, sheared raccoon, beaver or sheared rabbit requires more frequent cleaning than other furs. The fibers tend to become matted together by dampness, soil, or static electricity and must be restored by proper fur processing at least once a year. In large urban locales where air pollution of various kinds is more concentrated fur, as well as other fabrics and apparel, tends to fade, become dirty, and in general is subject to conditions that require cleaning more often. A coat that "has hung in the closet all year" may not require cleaning, although it will do it no harm.

Fur, like most items, looks better and lasts longer if cleaned when necessary. The furrier who suggests a yearly cleaning is not pushing it for the profits involved. There is little financial gain from giving this service. Ironically, he might be better off letting the garment go untended, because it will shorten its life and make a replacement necessary sooner. When you compare the total cost of cleaning a dress or suit against its original price, to the relative cost of cleaning your fur against its value, you are probably spending more percentage-wise on the cloth garments than on the fur. To conclude, most fur garments will look better and last longer if cleaned once a year. By all means, have it cleaned by someone who knows how to clean furs, along the lines described in prior chapters.

Q. 7   Shall I buy a fur-lined cloth coat?

Are you one of those who can't stand cold weather and is miserable whenever the temperature hovers around the freezing mark? Do you have to be out regularly in cold weather for any length of time? Do you live in an area where the temperature is often below zero and cold winds blow? If so, the fur-lined cloth coat, or even the fur-lined fur coat, is your answer. As warm as fur is when the hair is worn on the outside, it is much more so when turned inwards. The scientists tell us that the reason for this is the millions of small air pockets created by the fur fiber are warmed by the body heat and kept warm by the double barrier of leather and cloth which shields them and you from the cold.

Unfortunately, you can't have this kind of protection without bulk. It takes careful designing and choice of fur to keep the bulky look down to a minimum. The best skin for the inside linings should be warm, (amphibious again), light and thin. Lower grades of skins whose leather and hair are good, but do not have the desirable natural or dyed color required for regular fur-out garments are often used for this purpose. Nutrias, forms of muskrats, and fur seals are generally used, but others are suitable.

When a used fur garment has reached the end of its useful life as apparel and no longer has the looks and strength it once had, it can be converted into a fur-lined cloth coat. Such a conversion offers several advantages:

1. It adds a genuine storm coat to the wardrobe if and where needed.

2. It prolongs the useful life of the fur. The reasons for this are not too hard to understand. The cloth, not the fur leather, takes the full brunt of the strains of movement. Not all the fur of the original coat will be needed, so that the worst

parts can be cut away. Because most of the fur lining is inside and never shows, only a small percent of the old coat needs to be in fairly good condition to make a presentable and useful lining.

One of the common errors amateurs fall into is attempting to line a cloth coat with fur without professional help. The obstacle is not the extensive hand sewing, which is long and tedious, but the cut and fit. Furriers long ago learned at great cost that this work was a special hybrid of fabric and fur work best left to specialists who do nothing else. These conversions houses have learned how to select, adjust and allow for the bulk of the fur so that the garment hangs, looks, fits, and feels right.

Many department stores and specialty shops stock this item and many more contract with one of the specialists. They advertise a conversion sale, work with the customer in the selection, sales and measurements, and ship the coat to the specialist to be worked on. All in all, fur conversion offers an extended life to a still useful fur coat.

Q. 8   Shall I re-dye my faded or discolored fur coat?

Only in rare cases. It is not like re-dyeing a dress or curtain. Dyeing will not restore or hide worn hair. Indeed, it has a tendency to make the hair more brittle. Considerable labor is involved. The lining must be removed and all "closing" seams re-opened so that the garment lies flat. Rarely will the fur sections be dip-dyed, as with a dress. In most instances, the color is brushed on or *feathered* by hand, requiring the services of a skilled crafts-

man to produce the desired hue. Sometimes many successive light applications, plus spotting, are necessary to correct uneven fading. After the recoloring is completed, the garment must be completely remade. This includes nailing, resewing, squaring, and finishing. The leather of many older garments does not respond well to such treatment and is often weakened. If re-blending of recoloring is to be done at all, it is much less costly as part of a remodel since the garment needs to be flattened and relined anyway. The fees charged by the specialist colorists are high and must be passed on to the customer.

Q. 9   Shall I have my garment custom-fitted?

Yes, if the garment is in the luxury class. Of course, if you find a garment on the racks that fits you perfectly except for length of sleeves, body, or a button placement, by all means take it. This is more likely to happen when stocks are full or if you normally fit into a standard-sized garment without alterations.

Many a customer who will buy a dress off the racks will insist on a custom fitting of the gown for her wedding or other important occasion. A major investment in a fine fur coat falls into this category. The procedure is to decide on the style features that you want (which you can do when a custom garment is being made for you) and have a trial canvas fitting made up from your measurements. Sometimes a fur garment is available which has all the features you want, except that it doesn't quite fit you. This will serve in lieu of a canvas and give you a good idea of how the finished garment will look. Unquestionably, a custom-made garment will be more expensive than its stock duplicate, but it will be better in fit, individuality, and conformance to your individual physical and style needs.

Q. 10   How do I have a fur-lined cloth coat cleaned?

As if it were a fur coat. Dry cleaners will clean and ruin a fur coat by immersion, cleaning the hair beautifully but at the same time cleaning out the oils in the leather. The result is that the leather becomes brittle and begins to crack and tear. By all means use the services of a professional furrier.

Q. 11   How do I clean a cloth coat with an expensive fur collar or trim?

If at all possible, the collar or trim should be removed and cleaned separately as fur. All over dry cleaning is, however, less dangerous than for a fur lining because the leather of the fur trim is normally not subject to strain. The best procedure, slightly more expensive but worth it, is to have the whole coat cleaned by the furrier. He will do an excellent job on both.

Q. 12   I burned a small hole in my fur with a cigarette. How do I fix it?

Don't try it yourself. The burned area cannot be restored. It must be cut away and replaced with another piece of matching fur. This type of repairing can't be done by an amateur. Learn to be very careful of your precious fur in crowded places like elevators.

Q. 13   How can I sell my fur coat? How much can I get?

The answer to the last part of the question determines the rest. First, your garment is legally "second-hand" even if you sell it one hour after buying it. If you buy a new car, drive it around the block and bring it back for resale, you expect to lose several hundred dollars. Your best bet is to try to eliminate the middleman and find a buyer among your circle of friends and acquaintances. You can advertise, but that presents certain obvious problems including dealing with strangers.

The main difficulty is the wide disparity between what you expect to get for your garment "which still looks like new" to you and what you are offered by your furrier. He must take it in trade, and even the low figure he offers you gives him no opportunity for any profit by resale. Most reputable furriers will not sell second-hand garments directly to their own customers, because it tends to tarnish their business image and standing.

The best market for such used garments is the specialized second-hand or used-fur store. Some of these budget resale or thrift shops buy garments for direct retail resale. Others buy for export, shipping garments in quantity to South America and Europe. Neither of these outlets is in a position to offer you much for your garment. Their risks and overhead costs are such that they cannot afford to give you more than about one-third of the selling price that they hope to get, if and when they find a customer who likes the fur, style, lining, fit and price. This combination is increasingly rare as style changes in fur accelerate.

Q. 14   Shall I insure my fur coat?

By all means, if you can get it. You

may have difficulty getting insurance on furs, especially in certain large urban areas. You can usually have it added to your personal tenant's or owner's household policy. This will protect your garment against household fire and theft, but will not cover it off the premises. There are variations of policies which will protect you anywhere, but these are difficult to obtain. Most portable valuables like furs, jewelry, and cameras, are covered by a floater policy which simply indicates that it is insured on the move, wherever you go. Some people in large urban areas now have to place such all-risk policies with Lloyds of London at a rate of 5% a year. Check carefully with your insurance agent on all these questions. It is better to be thorough than sorry.

In calculating your insurance costs bear in mind that, like a car, the value of the garment depreciates each year. The following percentages are about the standard depreciations for each year:

> 2nd year—20%
> 3rd year—4%
> 4th year—10%
> 5th year—20%
> 6th year—25%
> 7th year—25%
> 8th year—25%
> 9th year—25%

In each case you figure on the evaluation left over after the previous year's reduction, not on the original price.

Q. 15   May I pin flowers or jewelry to my fur?

No, and no perfume either!

Q. 16   What style of pocket shall I select?

The one designed to go with your coat. The designer had a total look in mind. Slit pockets, patch pockets, horizontal slit pockets, will all bulge and wear with abuse. Don't overstuff. Don't keep your hands jamed into your pockets. Remember, wrist jewelry doesn't help either.

Q. 17   What are the best types of fasteners or closings for a fur coat?

The traditional hook and eye has the advantage of being light and unobtrusive, but as a unit it is not too effective for holding the coat closed in all positions. The button and loop combination has some "give" in the elastic loop, but the elastic tends to stretch or unravel when used. The button and buttonhole is the most elaborate, and is most common on double-breasted styles. This type of closing lasts longer than the other two, but is more expensive to repair and replace. Zipper closings, used on some sport and casual fur garments, suffer from the disadvantage of occasionally having the hair catch in the zipper, especially on the longer-haired furs. The "frog" or fancy passementerie closing unit is elaborately high-styled and is limited in use. On the other hand, its plebian cousin, the toggle and loop, is a favorite for ski and casual-wear fur garments. We should not neglect to mention the clutch-styled wrap which the wearer wraps around herself and holds in position by hand. Other novelty fastening systems such as snaps and pressure adhering surfaces have so far had limited success in furs. To conclude, none is perfect for all garments and every style. You will have to make a choice.

Q. 18   How do I take care of my furs at home?

The closet in which the fur is hung should be the coolest in the house. There should be plenty of room in the closet so that the fur is not crushed or stretched out of shape. Hang it on a wide-shouldered, shaped plastic or covered hanger. Be sure to center it carefully on the hanger so that the shoulder fits the shoulder. *DON'T* enclose the garment in an air tight plastic or cloth bag. On the contrary, make sure the air flows freely around it. If you don't wear it for a few weeks, take it out of the closet, hold it doubled with the front edges touching and shake it out vigorously. Periodically, hang it where it will get an airing, for a half hour or so, without getting it wet or being subjected to the direct rays of the sun.

Avoid airings in the spring, which is the flight period for several species of wild carpet beetles, when the females seek locations that will be good for a food supply for their eggs. If you happen to own a small portable fan or a type of vacuum cleaner which has an air blowing attachment, use it to blow out the fur as often as you can. Paradichlorbenzine blocks or nuggets normally used to protect woolens can be used for fur, but avoid direct contact of these chemicals with fur or wool surfaces.[1] Many people have an aversion to the persistent residue odor which lingers in garments stored with natural camphor.

When you wear your garment, re-member that durable as fur is if you wear an armful of bangles, your inside fur cuff and cuff edges will wear away quickly. Nor is your fur lining any stronger. Some of the most expensive all-silk linings do not last as long as those made from man-made combinations. If you are one whose body chemistry is such that your perspiration has a tendency to attack suit and dress materials, you can expect the same result on your fur lining. The addition of arm shields made from the same material as the lining will help maintain your fur garment lining. They can be easily replaced as they wear away. If possible, get some scraps of the lining for this purpose from your furrier when you buy the garment.

Q. 19   Why do my fur coat fasteners tear so often?

More buttons and fasteners are torn loose from fur garments than from any other type of outer garment, or so it seems to the average furrier. The wearer accustomed to the "give" and stretch built into the common cloth garment, forgets herself when wearing her fur and reaches for something in the fully buttoned garment. A fur garment is not constructed with that type of elasticity. Before sitting down, the wearer should be certain to open the bottom fastener and also gently lift the garment it avoid pressure when she sits.

Q. 20   Why is the seat of my coat worn?

Most fur-garbed women travel by car, plopping themselves down in a vinyl covered car seat. They may remember to open the fasteners but will sit on the coat, thus giving the fur every opportunity to be rubbed under pressure against the hard, abrasive car-seat material. Ample experience has shown that the car seat cover will win the abrasion contest by wearing away guard hair and fur fiber. What can you do? The best answer is to take the coat off or at least wear it lifted to the shoulders, folding it up in back so that you are not sitting on it. The longer the ride, the more important this becomes.

Q. 21   What do I do if my fur gets wet?

The oils implanted by nature in the fur fiber and leather will take care of any normal wetting. As soon as possible, shake it out vigorously and hang it carefully on a wide-shouldered hanger. Allow it to dry naturally in a cool dry place. Do not use heat in any form to assist the drying process. This would only singe the fur and dry out the leather. Do not allow your fur to become so wet that the dampness gets through to the fur leather and lining. This may create major problems of shrinkage, stretching, and require a visit to your furrier.

Q. 22   May I comb out my coat?

For all but the curly lamb skins a few strokes with a comb in the direction of the hair flow on those areas where the hair seems to be matted will do no harm. Fur hair and fiber wears away and tears loose if the pressure is exerted at the base of the fiber. Periodic light combings, a few strokes at a time, will help a little. Be careful not to do this too vigorously or to comb too deeply into the leather.

Q. 23   Why are garments made from the same fur as my coat advertised for so little in the newspapers, less than I paid?

Some of the reasons have been stated previously. The variation in the price of the best grade as compared to the poorest grade or section of any common fur in any season is at least one to four, if not more. At the same time when the finest persian lamb skin costs $25.00 each, it is not difficult to buy poor grades for $5.00. In some furs, the range is even greater.

Workmanship standards, alas, also vary. A fine fur manufacturing house is likely to spend at least three times as much as his popular-priced competitor on the lining and finishing of his garment, although both are made from the same fur. The same disparity holds true all through the steps in the manufacturing of the garment. It does not upset me to find some men's suits advertised for $49.00 while another ad has a price tag of $225.00. Both are wool, but when put side to side, or tried on, the difference is obvious. Why should furs be different, especially when the quality of material and labor can vary so greatly in furs, far more so than in material cut from a bolt? The picture shown in a newspaper ad does not have to be a picture of the garment being advertised at a low price. The customer is often shocked when she sees the garment being sold at the advertised price in the store.

Q. 24   When a store advertises 25% off and the like, shall I believe it?

Until the Federal Trade Commission

stepped into this situation, such claims of "savings" were uncontrolled. The question you should ask yourself is the same as for any other "reduction sale." Does previous experience show that sales in the store are reliable? What, if anything, is wrong with the item? Is it poor in quality, style, fit, or otherwise defective? The prospective buyer must determine the reason for the reduction and buy accordingly.

FOOTNOTE—APPENDIX C

1 Roy J. Pence, "Analyzing Fur Damage With A Microscope," University of California, Cir. 541, Oct. 1966, p. 17.

BIBLIOGRAPHY

Bibliography

Abbey, Edward, "Let Us Now Praise Mountain Lions," *Life Magazine*, New York, Mar. 13, 1970.

Adams, Arthur D., *Fur Bearers of North Dakota*, State Game and Fish Dept., Bismark, N.D., 1961.

Arizona Game and Fish Dept., *Bear*, 1968; *Pronghorn Antelope*, 1966; *Bighorn Sheep*, 1967; *Mountain Lion*, 1968; *Fur Bearers, Non-Game-Animals*, 1967; *Deer*, 1970; *Javelina*, 1969.

Ashbrook, Frank G., *Furs Glamourous and Practical*, D. Van Nostrand Co., New York, 1954.

Bachrach, Max, *Furs*, Prentice-Hall, Englewood Cliffs, New Jersey, 1947

Basch, H. and Co., *Manufacturing and Remodelling Persian Lamb*, Herman Basch and Co., New York, 1943.

Contini, Mila, *Fashion from Ancient Egypt to the Present Day*, Odyssey, New York, 1965.

Cronan, John M. and Albert Brooks, *The Mammals of Rhode Island*, Dept. of Natural Resources, Division of Conservation, Wildlife Pamphlet No. 6, Rhode Island, 1968.

Fouke Company, *The Inside Story of the Fur Seal*, The Fouke Co., Greenville, S.C., 1969.

Frankel, Jury, "Sables," from *Rauchwaren Handbuch* by same author, *International Fur Review*, London, May, 1969.

Fuchs, Victor, *The Economics of the Fur Industry*, Columbia University Press, New York, 1937.

Fur Age Weekly, *Classified Fur Source Directory*, Fur Age Weekly, New York, 1970.

Fur Information and Fashion Council, *A Fur Industry Report on Killing of Fur-Bearing Animals*, Fur Information and Fashion Council (Release), New York, 1970.

Fur Information and Fashion Council, *Report of Industry Leadership Conference*, Fur Information and Fashion Council, New York, Oct. 1966

Fur Information and Fashion Council (booklets).
Three Rules for Adding Years of Service to Your Furs, The Softest Touch, How to Put the Win in Windows, How to Build Sales Through the Mails, How to Make Small Space Advertising Sell for You, How to Reach New Customers with Radio, How to Sell Furs with Television, Picture Yourself in Furs, How to Sell Fabulous Furs, Fur Information and Fashion Council, 101 W. 30th St., New York, N.Y. 10001.

Fur Trade Review, *Photographic History of Furs*, Fur Trade Review, (Disc.), New York, 1917–18.

Hall, R. E. and Nelson, K. D., *Mammals of North America*, Ronald Press, New York, 1959.

Hildebrand, Oscar, *Principles of Fur Matching, Theory and Practice of Mink Dropping*, Souplesse Furs Ltd., London, 1966.

Kaplan, David G., *The Fur Book*, R. H. Donnelly Corp., New York, 1950.

Kaplan, David G., "Some Real Furs Look Fake," article in *Textile Services Management*, R. H. Donnelly Corp., New York, Mar., 1966.

Kybaloua, Ludmill and Others, *The Pictorial Encyclopedia of Fashion*, Crown, New York, 1968.

Latto, Ruth, "Alice in Fashionland," article in *American-Mitchell-Sol Vogel Magazine*, American-Mitchell Publishing Co., New York, July, 1948.

Leather Institute, "Glossary of Leather Terms," *International Fur Review* (quoting the Leather Institute), London, Sept., 1970.

Marden, Luis, *Titicacca, Abode of the Sun*, National Geographic Magazine, Feb., 1971.

Mech, David L. and Durward Allen, *The Wolf*, U.S. Bureau of Wildlife and Fisheries, Washington, D.C.

Mills, David C., "The International Fur Trade," articles in *Journal of Home Economics*, Nov., 1926, Dec., 1926, Jan., 1927, Feb., 1927.

National Education Television, *Our Vanishing Wilderness*, National Educational Television, Channel 13, New York, Oct. 11, 1970.

National Fire Protection Association, *Fur Storage, Fumigation and Cleaning*, National Fire Protection Association, No. 81, 69 Battery March St., Boston, Mass. 02110.

Neinholdt, Dr. Eva, "Furs in European Clothing," series of articles published in *Hermalin*, Hermalin Verlag, Leipzig, Germany, 1963–1966.

North Dakota Game and Fish Department, *Endangered Wild Life*, North Dakota Game and Fish Department (Release), 1970.

Pence, Roy J., *Analyzing Fur Damage with a Microscope*, University of California, Experimental Station Extension Service, Cir. 541, Calif., Oct. 1966.

Raphael, Samuel, *Advanced Fur Craftsmanship*, Fur Craftsmanship Publishers, New York, 1948.

Schöps, Dr. Paul, in cooperation with Dr. H. Braukhoff, Ing. A. Ginzel, Kurt Häse and Richard König, "The Durability of Fur," (trans. by Francis Weiss) for *International Fur Review*, London, June, 1965.

Sharpe, Ivy, "The Persian Lamb Story," *International Fur Review*, London, Sept., 1965.

Sharpe, Ivy, "The Progress of Beaver," *International Fur Review*, London, Feb., 1968.

Stringari, Fritz, Fiott and Burwell, *Petition to Federal Trade Commission on Behalf of Fur Dresser's Bureau of America*, Chicago, 1969.

Terrell, John Upton, *Furs by Astor*, William Morrow and Co., New York, 1963.

Terry, Frank, "The Muskrat," *Voice of the Trapper*, National Trappers' Association, Issue No. 3, Baltera, New York, 1969.

Thorer, Jurgen, "The Future of the Persian Lamb," *International Fur Review*, London, Mar., 1968.

Traianos, Anna, "Kastoria's Craft," *International Fur Review*, London, July, 1968.

Travis, Hugh F. and Schiable, Philip J., *Fundamentals of Mink Ranching*, Michigan State University, Dept. of Poultry Science, Circ. Bulletin, Michigan, 1914.

Tuttle, Terence, *How to Grade Furs*, Canada Dept. of Agriculture, Publication 1362, Ottawa, Canada, 1968.

Unknown, "Furs," *Fortune Magazine*, New York, Jan., 1936.

Unknown, *Furs and Their Romantic Past* (Pamphlet—author and publisher unknown). Approx. 1940.

U. S. Dept. of Interior, *Fur Seals of Alaska's Pribilof Islands*, Fish and Wildlife Service, Washington, D. C.

U. S. Dept. of Interior, *State Wildlife Agencies*, Office of Conservation Education, Bureau of Sport Fisheries and Wildlife, 435R, Washington, D. C., Nov., 1969.

U. S. Federal Register, Vol. 35, No. 106, Washington, D. C., Tuesday, June 2, 1970.

Various, *What Do You Know About Furs?*, series of articles by various specialists which appeared regularly on Wednesdays on Fur Page of *Women's Wear Daily*, Fairchild, New York, 1938–1942.

Weaver, Mark, "Future of the Beaver," *International Fur Review*, London, May, 1969.

Weiss, Francis, *Adam to Madam* (unpublished manuscript, partly published as a series in), *International Fur Review*, London, 1963–1966.

Weiss, Francis, J. L. Debourciev, Judith King, Fred W. Meiser and Johan A. Strand, "Glossary of Seals," *International Fur Review*, London, July, 1964.

Wilcox, R. Turner, *The Mode in Furs*, Chas. Scribner's Sons, New York, 1951.

Women's Wear Daily, *Sixty Years of Fashion—1900–1960*, Fairchild, New York, 1963.

Woolsey, Norman, "Voice of the Trapper," *Arizona Report*, National Trappers' Association, Baltera, New York, Mar., 1970.

INDEX

Index

pony, 186–7
Pope Paul VI, 10
porpura shell, 3
preparing and stretching, 71–2
printing, 38
processing, 14, 25–6, 30, 31, 32–3, 36–41, 69–87, 77 (chart); coloring, 38–40; cutting, 72–81, 86; dressing, 36–41; innovations in, 86–7; plucking and shearing, 40–1; recent trends in, 138–41; skin sorting, 70–1; stretching and preparing, 71–2
puma, 53

Q

Quality Manos Block, Inc., ix
questions and answers about furs, 211–19

R

Rabbit, 4, 5, 11, 21, 37, 41, 60, 76, 100, 187–8, 187 (illus), 212; volcano rabbit, as endangered species, 200
"Rabid Foxes Rampant Locally," 132f
raccoon, 17, 75, 120, 130, 154, 188 (illus), 212; American raccoon, 188–9
ranching, 26–8, 29–30, 32, 70, 77 (chart); and genetics, 26; ranchers' associations, 28, 35, 77 (chart); stock expert, 27–8; stock smuggling, 27; volume of, 135
"Record Prices of Pelts" (Haber), 132f
Red Data Book, 120
remodelling, 97 (illus), 99–103; decision to remodel, 100; expense of, 99–100; fitting, 102–3, 214; merchandising remodels, 102; pattern selection, 101; and types of fur, 100
Renaissance, 9, 11–12
repairing, 95–7, 95 (illus), 98–9; bald spots, 96–7; fabric tear, 95, 96; fastener replacement, 95–6; open seam, 96; pocket edge tear, 96; re-edging, 98–9
Rhode Island, regulations affecting wildlife, 207
Richlieu, Cardinal, 5
Romary, Fred, ix
Rome, ancient, 3, 4, 9
rugs, fur, 109 (illus)
Rupert, Prince of England, 5

S

SAGA, 28, 148
SWAKARA, 28, 148
sable, vii, 4, 5, 12, 30, 59, 189–90, 189 (illus)
"Sables" (Frankel), 197f
Samet, Arthur, 197
Scandinavia, as fur center, 27–8, 35, 36
Scandinavian Mink Ranchers, 28
scarf, fur, 11, 17, 59–60
Schreibman-Raphael, Inc., ix
Scops, Paul, 197f
seal, 3, 13, 16, 25, 30–1, 35, 40, 59, 76, 83, 100, 117, 130, 138, 190–2, 212, 213; Alaska fur seal, 4, 30, 31, 59, 138, 190–1, 190 (illus), 211, 212; cape seal, 31; Guadaloupe fur seal, as endangered species, 199; hair seal, 191–2, 191 (illus); Hudson seal, 30, 40; Lakoda seal, 138; killing of baby, 121, 125; protection of, 29, 117, 124; substitutes for, 30–1
Semiramis, Queen of Babylon, 3–4
sewing, 78, 80 (illus), 87 (illus), 95 (illus)
shako, 12
Sharpe, Ivy, 41f
shearing, 140
shearling, 58
"Shearling Look Makes Ski Scene," 197f
sheepskin. *See* lamb
Shenker College of Fashion and Textile Technology (Israel), 49
Shooting Times (magazine), 126
shrug, 60
Sidney Lambert, Inc., ix
silk screening technique, 138–41
skin joining, 75–83, 76 (illus); and block layouts, 75, 75 (illus); and assembled furs, 76–8; and concealed seams, 75–6; flowering, 76; 'let-out' technique, 78–81; nailing, 81–3, 87–8, 214; sewing, 78, 80 (illus), 87 (illus), 95 (illus)
skinning, 25
skins, 32–3; dealers, 69–70; match- 71 (illus); defined, 25; mink, 34 (illus), 36 (illus); sorting and packaging, 70–1; stencilling, 138–41; stretching and preparing, 71–2
skirts, fur-trimmed, 15 (illus)
skunk, 192

W

Wallaby, 194–5

Washington State, regulations affecting wildlife, 208

Wasserman Sports, 26

Watteau and DeLille, 13

weasel, 195

Weiss, Francis, 5f, 22f

West, Mae, 18 (illus)

West Virginia, regulations affecting wildlife, 208

wildlife protection, 28–9, 117–32

Windsor, Duchess of, 167

A. Winick and Son, ix

Winnipeg, as fur center, 33

Wisconsin, regulations affecting wildlife, 209

wolf, 153, 195–6; Eastern timber wolf, as endangered species, 199; timber wolf, 153; Texas red wolf, as endangered species, 199

wolverine, 153, 196

wombat, 154

Women's Wear Daily, vii, 117, 132f

woodchuck, 154

Wyoming, regulations affecting wildlife, 209

Y

Youth market, 21–2, 52–3

Z

Zebra, 21, 53, 59, 107, 138, 140, 154, 196

zebu, 53, 154. *See also* buffalo

ABOUT THE AUTHOR

About the Author

David Gordon Kaplan, author of *World of Furs*, was born into the fur industry, and he never left it.

Like his father, Mr. Kaplan has been both fur union member and businessman. But his involvement in furs has been much longer and deeper.

Among other interests, Mr. Kaplan has been an author, wholesaler, retailer, resident buyer, consultant, educator, and guest lecturer.

His range of fur activities is most creditable:

- He has taught war veterans at the Sol Vogel School, deprived students at the High School of Fashion Industries, and the well-to-do at Tobe-Coburn.

- He has written hundreds of articles on furs for publications in the U.S., Great Britain, Scandinavia—wherever fur interest existed.

- He has been consultant to the New York City Department of Markets, U.S. Treasury Bureau of Customs and, among many others, the Department of Industry & Commerce, Province of Manitoba.

Mr. Kaplan wrote *The Fur Book* in 1950, but never stopped collecting data on what he considered the ultimate book, an ambitious tome that would incorporate, for both layman and professional, all the vital elements of an ancient industry.

While working on *World of Furs*, he retained his business association with his former students, the Antonovich Brothers, and also continued column-writing and lecturing.

Under doctor's orders, he has had to cut back on activity. But his enthusiasm for the fur industry remains as keen as ever. Maybe more so, with the realization, after four years of labor, of *World of Furs*.